More praise for

THE HIDDEN HALF OF

"An ingenious idea! *The Hidden Half of Nature* draws a straight line from the microbes that live in healthy soil to those that live in healthy guts, skillfully blending the personal and the scientific. This is a must-read for anyone concerned with their own health." —AMY STEWART, author of *The Drunken Botanist*

"A stunningly clever book that connects the tiny dots of the microbial world. I will never ignore my microbes again."
—KIRK JOHNSON, Sant Director, Smithsonian National Museum of Natural History

"Beautifully written and engaging, *The Hidden Half of Nature* shows us the underlying principle of all life on earth—the microbes that live in, on, and around us. Montgomery and Biklé connect the strands with ease and consummate skill." —MARTIN J. BLASER, MD, author of *Missing Microbes*

"Hidden no more! Montgomery and Biklé give microbes their rightful due as powerful organic machines propelling life into death into life. I found myself shockingly entertained by their exposé of the 'wee beasties' of the microbial world." —BRUCE BARCOTT, author of *The Last Flight of the Scarlet Macaw*

"I love this book! It's genial, erudite and wise. Using their personal story, historical fact and cutting edge science, Montgomery and Biklé have given us a great gift—a deep understanding and appreciation of our relationship with the microbial world. Seeing this invisible world in an entirely new way, as allies, we can reclaim the health of our bodies, soil, and world. It's exciting, absolutely necessary, and, now that I have digested this wonderful work, I believe totally possible." —DEBORAH KOONS GARCIA, filmmaker, *The Future of Food* and *Symphony of the Soil*

"An absolutely wonderful book—one of those rare books that will truly deepen your understanding of the world. Combining personal narrative and rigorous science, and ranging from geology to horticulture to gastroenterology, *The Hidden Half of Nature* reveals the countless wonders of microbial life. I wish I had learned this in medical school!"

—HOWARD FRUMKIN,
dean, School of Public Health, University of Washington

"Microbes and soil. Microbes and humans. With a full sweep of knowledge, Montgomery and Biklé help heal the split in the ancient subject-object dualism."

—WES JACKSON,
founder and president of the Land Institute

"In a fascinating narrative explaining the economics and theatrics of microbes, Montgomery and Biklé bring the unseen world to life like great storytellers."

—JOEL SALATIN, Polyface Farm

"A much-needed microbial manifesto that unearths the mysteries and wonder of microbes and the intimate links among the health of the soil, the planet, and our personal health."

—KATHIE MADONNA SWIFT,
nutritionist and coauthor of *The Swift Diet*

"A wonderful read! *The Hidden Half of Nature* not only explores the workings of the microbial world, but also the people and the methods used to advance our knowledge of it."

—JEFF LOWENFELS,
author of the *Teaming With* series

"What a friendly and highly readable excursion. Through personal experiences and lively historical vignettes, the authors lead us on a delightful journey to discover why microbes are essential for the health of all life forms—including us."

—MOSELIO SCHAECHTER,
San Diego State University and University of California at San Diego

"This is a very timely book. Environmentalists and the public focus on and respond to megafauna—the big charismatic creatures from elephants and tigers to whooping cranes and monarch butterflies. But we forget that for most of this time that life has occupied Earth, it's been a microbial world."

—DAVID SUZUKI, host of CBC's *The Nature of Things*

"*The Hidden Half of Nature* offers a wonderfully fresh and exquisitely informed approach that could change how we relate to our selves, our diets, our gardens and our world." —TIM McNULTY, *Seattle Times*

"The science throughout is amazingly detailed and well-researched. . . . The implications for the health of our planet and the health of human beings are powerfully stated. . . . [*The Hidden Half of Nature*] skillfully weaves Biklé's cancer and health issues and lays out the beautiful connection between the microbial garden in our bodies and the microbial garden in the Earth." —SALLY PETERSON, *Oregonian*

"Written in clear and accessible prose, [*The Hidden Half of Nature*] has important messages for research scientists, physicians and the rest of us as well." —*BUFFALO NEWS*

"One of the year's best books on gardens and health."
—JIM McCAUSLAND, *Sunset*

"Even if you've never entertained a single thought about 'the parallel roles of microbes in maintaining the health of plants and people,' this book is an accessible and illuminating read." —*SEATTLE WEEKLY*

"In an accessible book on the new microbiology, David R. Montgomery and Anne Biklé . . . offer the twists and turns of their married lives in Seattle to illustrate their insights into the invisible. . . . [T]horoughly readable."
—NICHOLAS MONEY, *Wall Street Journal*

"Exceptionally inspiring." —FRED KIRSCHENMANN,
Leopold Center for Sustainable Agriculture

"The weaving together of the two strains of inquiry—the history of our thinking about microbes in soil and about those in guts—presents a different, coherent way of looking at both." —CELIA STOREY,
Arkansas Online

"The information is worthwhile, and suggests a transformational reconsideration of the status quo in agriculture, medicine, and nutrition."
—BARBARA LLOYD McMICHAEL, *Bellingham Herald*

Also by David R. Montgomery

Growing a Revolution: Bringing Our Soil Back to Life

The Rocks Don't Lie: A Geologist Investigates Noah's Flood

Dirt: The Erosion of Civilizations

King of Fish: The Thousand-Year Run of Salmon

THE HIDDEN HALF OF NATURE

The Microbial Roots of Life and Health

DAVID R. MONTGOMERY

AND ANNE BIKLÉ

W. W. Norton & Company

Independent Publishers Since 1923

New York London

Printed in the United States of America
First published as a Norton paperback 2016

For information about special discounts for bulk purchases, please contact
W. W. Norton Special Sales at specialsales@wwnorton.com or 800-233-4830

Manufacturing by LSC Harrisonburg
Book design by Kristen Bearse
Production manager: Julia Druskin

Library of Congress Cataloging-in-Publication Data

Montgomery, David R., 1961– , author.
The hidden half of nature : the microbial roots of life and health /
David R. Montgomery and Anne Biklé. — First edition.
p. ; cm.
Includes bibliographical references and index.
ISBN 978-0-393-24440-3 (hardcover)
I. Biklé, Anne, author. II. Title.
[DNLM: 1. Microbiota—physiology. 2. Conservation of Natural Resources.
3. Crops, Agricultural. 4. Health Promotion. 5. Soil Microbiology. QW 4]
QR111
579'.1757—dc23
2015027979

ISBN 978-0-393-35337-2 pbk.

W. W. Norton & Company, Inc.
500 Fifth Avenue, New York, N.Y. 10110
www.wwnorton.com

W. W. Norton & Company Ltd.
15 Carlisle Street, London W1D 3BS

5 6 7 8 9 0

E pluribus unum—
Out of many, one.

Contents

INTRODUCTION

We are living through a scientific revolution as illuminating as the discovery that Earth orbits the Sun. Yet the current revolution does not focus on massive planetary bodies, but on life too small to see with the naked eye. An explosion of discoveries is rapidly unfolding about the life below ground, within us, and literally everywhere on the planet. What scientists are finding is that the world we know is built upon a world we have mostly overlooked.

Throughout history, naturalists relied on their unaided eyes, ears, and hands to uncover nature's secrets. But our senses handicap us when it comes to her hidden half, leaving the microscopic world shrouded in mystery. Only recently have new gene-sequencing technologies and more powerful microscopes given us a window into this world. Scientists are now seeing that complex microbial communities drive many things we depend upon, from soil fertility to a healthy immune system.

Our curiosity about the foundational importance of microbial ecology arose in the unlikeliest of ways—on a stay-at-home journey to our backyard. Both Anne and I had been trained to observe and notice nature. I through my study of the geological forces that have shaped Earth's landscapes over immense periods of time, and Anne in her work as a biologist and an environmental planner in the field of public health.

So when we bought a house, began to dig into the soil of our new yard, and discovered we had fixer-upper dirt, our professional training kicked into gear. At the front of our minds was figuring out how our lousy soil would ever support a garden. Anne acted first, by following her gut and feeding organic matter to our lifeless soil. A *lot* of organic matter. Ton after ton of coffee grounds, wood chips, leaves, and home-brewed compost tea disappeared into the ground. Lo and behold, in what seemed like no time at all, the plants in our new garden began to thrive and grow like mad.

I didn't see how something as dead and unappealing as the organic matter Anne added to our soil could so rapidly spawn a blossoming of life. It was this simple mystery that launched us both on a mission of discovery. As we soon learned, microscopic soil life chewed and chomped its way through the organic matter, turning it into a smorgasbord of nutrients for our emerging plants. The idea of another world run by tiny, invisible, and largely unknown creatures fascinated us. Yet their undeniable effect rippled upward from beneath our feet. In less than a decade our yard went from a barren lot to a garden teeming with life.

Watching the amazing impact of our renewed soil showed us a solution for one of the oldest problems plaguing humanity—how to grow food without depleting or destroying the soil. The experiment unfolding outside our back door confirmed the pioneering insights of early organic farmers and gardeners. By nurturing the microbial life below ground, we can reverse much of the damage caused by the ancient practice of plowing and the modern overuse of pesticides and fertilizers. But our journey didn't stop there. We found that the life of the soil is only part of the hidden half of nature.

When Anne was diagnosed with cancer, it led us to question the notion of health itself. Where does it come from? This was when our view of the microbial life within us began to change. At first we shared the conventional medical view that sees microbes mostly as pathogens. We had both experienced the power of modern medicine to combat infections and were thankful for the antibiotics that had saved our lives. But it is not only the bad actors of the microbial world that influence our health.

The latest revelations about microorganisms show that we are not who we thought we were. This was brought into sharp relief a few years ago when a large consortium of scientists reported their findings in the journals *Science* and *Nature*. An unfathomably vast array of invisible life—bacteria, protists, archaea, and fungi—thrives on us and in us, as do innumerable viruses (which are not considered alive). Their cells outnumber our own cells by at least three to one, and many say ten to one, yet we are only beginning to learn what they do for us. And our planet—like the bodies of plants, animals, and people—is literally covered, inside and out, with microorganisms. Not only are they abundant, they're robust, able to withstand the most extreme conditions the planet can offer. To help you navigate the exciting world of microbes, we've included a glossary, a few endnotes, and an extensive list of sources for those who may wish to delve deeper.

The more we looked into these recent discoveries, the more we were intrigued by the parallel roles of microbes in maintaining the health of plants and people. And we learned the new name for microbes that live on us and in us—the human microbiome. We began to see how microbes could help restore soil fertility and counter the plague of modern chronic diseases. We had stumbled upon a whole new way of seeing nature.

In this book we tell the story of our journey uncovering and connecting ideas and insights about the emerging revolution swirling around nature's hidden half. We lean on, draw from, and champion the work of dozens of scientists, farmers, gardeners, doctors, journalists, and authors. It is a story that explores humanity's relationship with microbes. We are now realizing that microbes, long seen as invisible scourges, can help address some of the most pressing problems facing us today.

This new view of microbes is shocking—they are essential parts of us and plants, and always have been. Such a view points to an astounding potential for promising new practices in agriculture and medicine. Think animal husbandry or gardening on a microscopic scale. By cultivating beneficial soil microorganisms on farms and in gardens, we can ward off pests and boost harvests. And in medicine, research on the microbial ecology of the human body is driving new therapies and treatments. A few decades ago such ideas would have sounded as preposterous as invisible life itself did a few centuries before that. The emerging science of a microbial basis for health directly challenges the wisdom of indiscriminant campaigns against microbes in agricultural soils and our own bodies. Some are our secret silent partners.

There is no doubt that studying the natural world in neatly compartmentalized subjects lets us grasp the otherwise incomprehensible complexity of the whole. Specialization has allowed scientists to chalk up spectacular successes and discoveries. This is the standard approach in seeking cures and treatments for what ails crops and people. But this limited vantage point conceals broad connections fundamental to the microscopic world and our own.

It doesn't help that profound changes have occurred in the way scientists write about and communicate their scientific discoveries. Pick up a copy of *Science* or *Nature* from a century ago, and the average reader can understand what the authors of pretty much any article are talking about. Not so today. Modern scientific jargon is, for the most part, dumbfoundingly mind-numbing. Not to pick on any particular research group or journal, but in researching this book, we often found ourselves wading through sentences like this:

> *Recognition of peptidoglycan by NOD1 in IECs elicits production of CCL20 and b-defensin 3 that direct the recruitment of B cells to LTi-dendritic-cell clusters in cryptopatches to induce the expression of sIgA.*[1]

Unintelligible to most, this is actually an example of succinct scientific writing—the kind that advisors and editors encourage, and sometimes insist on. It packs a page into a sentence. But who, other than technical specialists in that field, can comprehend its meaning? In simpler terms this phrase says that certain intestinal cells recognize particular types of bacteria, and that this bacterial recognition causes immune cells to release substances critical to health. Of course, it conveys more details, like the name of the particular molecules and immune cells involved. But sometimes clarity on the specifics can obscure larger messages. And the more we delved into microbiome science, the clearer it became that we all need to know far more about how microbial ecology affects our well-being and our environment.

Researchers in microbiology and medicine are uncovering the intricate symbiotic relationships that exist between people and the microbes living in and on our bodies. Bacterial cells live alongside the cells lining our gut, where, deep within our bowels, they teach and train immune cells to sort friend from foe. Likewise, soil ecologists have made strikingly similar discoveries about the effects of soil life on plant health. Bacterial communities inside of and around plant roots help sound the alarm and man the barricades when pathogens storm the botanical gates.

As it turns out, the vast majority of bacteria in the soil and in our bodies benefit us. And throughout the history of life on land, microbes repeatedly deconstructed every piece of organic matter on the planet—leaves, branches, and bones—fashioning new life from the dead. Yet our relationship with the hidden half of nature remains modeled on killing it, rather than understanding and fostering its beneficial aspects. In waging war against microbes for the last century, we've managed to unwittingly chisel away much of the foundation on which we stand.

And while impressive and transformative new products and microbial therapies are on the horizon for both agriculture and medicine, there is a profoundly simple reason we should care about the hidden half of nature. It is a part of us, not apart from us. Microbes drive our health from inside our bodies. Their metabolic by-products form essential cogs of our biology. And the tiniest creatures on Earth forged long-running partnerships with all multicellular life in the evolutionary fires of deep time. All around us they literally run the world, from extracting nutrients plants need from rocks, to catalyzing the global carbon and nitrogen cycles that keep the wheel of life turning.

It's time to recognize the essential roles microbes play in our lives. They shaped our past and how we treat them will shape our future in ways we are only beginning to understand. For we will never escape our microbial cradle. Nature's hidden half is as deeply embedded in us as we are in her.

1

DEAD DIRT

We didn't set out to learn anything. Anne wanted a garden and I didn't object. Neither of us suspected that buying an old house on a ramshackle lot in north Seattle would change how we saw ourselves and the world. After all, the yard wasn't anything special—a decrepit, century-old lawn hosting a few plants blown in on the wind. Like most homebuyers, we didn't bother to look under the hood at the soil beneath. Just a century ago old-growth forest covered our lot. Surely, we assumed, it could support a garden. So we hired an inspector to check out the house, and didn't give the soil a second thought.

Several years later, and after much planning, the long-awaited day finally arrived. We stood outside on our freshly cleared, bare lot in baking sunshine surrounded by dozens of potted shrubs and trees. They looked like trained dogs patiently waiting for dinner. We were ecstatic and anxious. Early August is a dicey time to put plants in the ground. But after construction snafus and delays that ate up the spring, this was where we landed.

Anne grabbed the shovel handle with both hands and jumped onto the ledge of the blade. It slid slow and smooth into the ground. She looked up at me, beaming. Her dream of a garden was becoming real. She piled up the soil next to the hole. It was medium brown and still a touch moist from June rains. On her third jump a dull *t-i-i-i-n-g* rang up from the hole and her feet bounced off the shovel. She tried again. *T-i-i-i-n-g!* She hurled the shovel into the hole like a javelin. It hit the bottom and fell aside with a thud.

We both peered down into the six-inch hole. A thin milk-chocolate layer at the top gave way to swirling khaki and taupe. Beneath, the soil took on a greasy sheen, and its color changed to mouse-gray shot through with streaks of beige. The problem lay in plain sight at the bottom of the hole—marble- to golf-ball-sized rocks tightly packed in hard clay. Glacial till. I looked up at Anne and saw panic in her eyes as she scanned the plants slowly cooking in their black plastic pots scattered across the bare dirt of our lot. We'd planned this day for years and now the damn dirt wasn't cooperating.

How could this have happened to us? Frankly, it's a little embarrassing. As a geologist, dirt and rocks are my domain. And, as a biologist turned gardener, plants are Anne's territory. Who could be better positioned to notice and appreciate the quality of soil than a pair like us? Soil straddles our worlds, and yet we each saw only half the story.

In that moment I regretted not having thought to dig a soil pit in the yard before we bought the house. In my fieldwork all around the world, I'd dug plenty of holes to see what lay beneath my feet. But somehow it had not occurred to me to look below the surface of a city lot, of *my* city lot. So there we stood wondering how to turn rock-hard till into a dream garden. Whatever the answer, Anne insisted the plants could not remain in their pots any longer. So into the ground they went.

Nature's geologic smackdown created our predicament 17,000 years ago when a glacier surged south out of Canada and overran Seattle and much of western Washington. Because this ancient mountain of ice had steamrollered our lot, we now faced a twofold problem—a miserly dusting of organic matter over weathered clay and concrete-like glacial till. Short of digging up our entire lot and hauling in new dirt, we couldn't change the hardpan till. That would have to stay. But we could do something with what lay closer to the surface.

Fertile soil is the frontier between geology and biology, a mix of weathered rock fragments and decaying organic matter. We had the rock part of the recipe but not the other part. We needed organic matter, *a lot of it.*

We both knew that dump-truck loads of commercial compost would further bust our already-blown budget. And I pushed for low-smell, as well as low-cost, ideas, recalling the time Anne had loaded up my pickup truck with chicken manure and dumped it all over the front yard of the place we were renting. It stank for weeks. I was in no hurry to repeat that experience. While I could see that boosting the paltry amount of organic matter in our soil was a good idea, I didn't want Anne to turn our new yard into a compost heap.

So, instead, she came up with an idea I could hardly object to. There would be no smell, no effort on my part, and no expense—zip, nada, nothing. She wanted to start looking around for wood chips and leaves and pile them on top of our new planting beds. While this was fine with me, I didn't find the idea of a mulch fest very impressive. What real impact could bits of wood and dead leaves have? I thought the wood would take forever to break down and the leaves would blow away before they rotted and mixed with the soil.

Still, I could see that Anne's mulch scheme did offer an immediate advantage. She had spent much of August and September running around, sometimes twice a day, with watering cans, sprinklers, and hoses heaping fresh curses on the landscape architect for his repeated schedule delays. If she covered the soil with wood chips, it would help keep moisture in the ground and tame our skyrocketing water bill.

Anne's hard work kept at bay another problem, something that terrified her inner gardener. She'd seen it on other people's lots and yards all around town; only the tony homes with armies of gardeners seemed exempt. She'd seen it at work too, watching the botanical scourge of invasive plants spread up and down the Cedar River, threatening newly planted native species in restored floodplains. With Seattle's mild, wet climate, unwanted plants muscle and bully their way into gardens and natural areas. Our house came with a charter member of this undesirable club—Japanese knotweed. It's the kudzu of the Pacific Northwest, spreading like wildfire and just as hard to get rid of. When we bought the house, a big clump stood like a concierge at the foot of the driveway. Anne cheered when the bulldozer operator doing site prep for the garden took it out in one deft move and tossed it into a dump truck. She knew that diligent mulching would help keep uninvited guests out of our new garden.

Her first score of organic matter was from the neighborhood. She had heard that arborists would give away wood chips to avoid paying disposal fees. So she opened the phone book and started calling. Ballard Tree Service, Seattle Tree Preservation, and more—getting her name on their drop-off lists. After a while, when a crew was on a job in the neighborhood and their truck got full, they'd stop by and dump five to ten yards of wood chips all over our driveway. Then Anne made friends with a neighbor who had a corner lot with almost a dozen oak trees on the planting strip. When fall rolled around, he was perfectly happy to let her haul away loads of dropped leaves. She developed a knack for spotting unwanted organic matter—like the bags of espresso grounds that she snagged from behind the neighborhood coffee shop, or any of the legion of Starbucks around town.

The sorry state of our lot reflected the priorities of the two previous owners. A Norwegian family, the Osterbergs, built our house in 1918. Judging from a 1930s assessor's photo, they weren't gardeners. The dim black and white picture, taken from the sidewalk, looks up the driveway hugging the east side of the house to the garage at the back of the lot. A sea of patchy, rough grass surrounds the house, a pile of scrap wood leans against the garage, and a low, flimsy chicken-wire fence encircles the front yard. In 1988, the last Osterberg died and a guy who'd grown up in the neighborhood bought the place. Gardening apparently held little interest for him as well. When we bought the house, the lot looked remarkably similar to the 1930s photo, less the chicken-wire fence.

And yet I really didn't think it was all that bad. The eighty-year-old lawn of weeds gave the place a green veneer and was ideal for throwing tennis balls for Xena, the Lab-Chow mix we had liberated from the city pound. The grass took no care and would spring back to life with the first autumn rains. Anne, however, considered the yard one big problem from the get-go, and I was content to let her plan the garden. Actually, I knew there was no stopping her. She wanted a garden. Bad. And neither of us wanted me in charge. Anything green entrusted to my care never lasted more than a few months.

Foremost in her mind was figuring out how to convert a sea of grass clumps and old-growth dandelions covering the hundred-foot-long side yard east of the house into a garden. A small patch of the same lay in front and back of the house, giving us a stretched-out U-shape to work with. In the south-facing back yard a Douglas fir and a holly were intertwined like circus freaks, and a King Kong–like horse chestnut throttled the corner of the house. Brambly deciduous shrubs along the western property line were smothering one another in a fight over light and space.

In her mind's eye, Anne imagined a garden filled with greenery that spilled out to the street. She wanted leaves and petals unfurling every season. A garden, she thought, should be a place that would enchant us and turn our heads every day. How to make such a place, I would soon learn, is as much about the art and craft of cultivating hidden soil life as it is about tending the plants we admire.

Littleton, Colorado, the suburb south of Denver where Anne grew up, isn't really the kind of place that breeds plant lust. Back then, the Denver area was better known for its stinky stockyards and the brown haze of wintertime air inversions than for bucolic gardens. With no delphinium-doting aunt or grandmother to take her under her wing and teach her the ropes of gardening, Anne sprouted a green thumb anyway.

She loved to watch Colorado's seasons smash into one another. When she was about seven, she first noticed the jewel-like orbs that burst from beneath March snows in a small garden next to the driveway. The gutsiness of cherry-red tulips pushing up into a frozen, white-draped world impressed her. So did the irises that appeared every June in a small isolated rock garden on its way to being overgrown. Their flowers magically emerged from knobby, dead-looking rhizomes exposed between rocks and other bigger plants. On tiptoes, she'd stick her nose as far as she could inside the purple flowers to feel their velvet petals brush her small face. She breathed in deep gulps, their fragrance like grape soda. To this day, she can't resist big, gaudy irises.

When Anne was a teenager, she filled an old, round wicker table in front of a south-facing living-room window with plants. Philodendrons threaded their way between pots and silver-spotted begonias towered over tiny cacti. She played God, moving plants closer to and farther from the window to watch how they reacted. Their phototropism intrigued her. She could make plants bend and twist and return to upright in a matter of days. Over the years she developed a knack for pulling greenery back from the edge of death. Plants could lift her mood, awe and inspire her—plants possessed power.

Although we grew up in different places, we both did similar things. Like most kids who grew up in 1960s and '70s suburbia, we played outside in the yards, vacant lots, and empty fields of our sprawling neighborhoods. In the summer we mounted expeditions to nearby fields and creeks, catching whatever we could outsmart to bring home in a can or a jar to watch for a while. These places held pockets of nature that shaped our lives and were easy to touch, see, and smell.

Raised in California, I loved the outdoors. My interest in geology began on hiking trips in the Sierra Nevada with my Boy Scout troop. Traversing the state's granite spine gave me a sense of the land as a living system and honed my map-reading skills. These early experiences, contemplating the shapes of landforms, eventually led me to the world of geomorphology, the study of Earth's dynamic surface.

When I was a teenager, my bedroom looked toward "Cow Hill," the last part of the Stanford campus that served as a working farm. My brother, our friends, and I would roam through fields and gullies, hide in the tall grass, and climb gnarled oak trees. Halfway through college, I took time off to work in an Australian mine and travel the South Pacific. In faraway places, miles from any city, I found real nature. I lived among kangaroos and saltwater crocodiles in the Outback, learned to dive on the Great Barrier Reef, and explored the Southern Alps of New Zealand. These experiences cemented my ideas of what nature was and where I could find it.

Anne's interest in nature grew on family camping trips deep into the Rocky Mountains and blossomed into a passion for biology. When she moved to Santa Cruz for college, she was shocked. Here on the central coast of California, plants grew riotously, as though in perpetual spring. This only watered her plant lust, which blossomed later in graduate school, where we met.

She and her five housemates tore out the sod from their postage-stamp-sized yard in North Berkeley and built a little picket fence along the sidewalk in the south-facing front yard. They planted a tree, a dozen flowering perennials, and several shrubs. Anne became chief caretaker. She kept an eye out for which plants needed the most water or protection from the rare freeze, using fallen leaves and cut-up plant debris to blanket the soil. I liked their garden and the little brick path you walked along to reach their front porch. Anne's touch made it inviting and wild at the same time.

After I finished my Ph.D. and was hired at the University of Washington, we moved to Seattle. Anne talked our new landlord into letting her put a couple of small vegetable beds in the front yard of the place we rented. This was when I learned about the stench of chicken manure, and Anne learned of its power to spur plant growth. In her pair of small, chickenshit-infused vegetable beds, she grew bumper crops of tomatoes, beets, lettuce, and basil. We hardly had to buy produce all summer.

When we finally bought our house, Anne started laying plans for a *real* garden. We both liked the idea of bringing more nature into our daily lives. But Anne's gardening experience didn't make it any easier to design the garden of our new home. She'd never had so much space. More than that, our yard was anchorless. There was no hint of a once-glorious garden to build on, no crumbling stone wall or forgotten gate covered in flowering vines, no languishing perennials hidden in weed-infested beds.

So Anne turned to books and magazines filled with information and pictures about patios, fences, plants, and more. She relied on Seattle's neighborhood garden tours, where you could visit a dozen or more gardens in a day. She poked around the featured yards, peppering the owners with questions, and took a lot of pictures. Back home, she cut up the photos and made collages. I thought they looked like oversized ransom notes from a crazed botanist. But they helped us imagine what the garden could look like and how we might use it.

Anne researched the size and shape of trees and shrubs, and whether they could withstand Seattle growing conditions—ferocious winds and waterlogged soil in the winter and drought in the summer. Were they sun lovers or haters, workhorses or racehorses in terms of growth? Which would be the most beautiful or interesting? In what seasons? What about in twenty years?

We began to imagine how trees and shrubs could screen our yard from the street and our immediate neighbors. A vision came into focus. Out of sight meant out of mind. We'd fill a planting bed with trees along the east side of the yard, and plant beds with fewer trees and more perennials on the south and west sides.

We'd banish our cars to the street and recast the driveway to our small detached garage at the rear of the lot as a path through the garden. We always came and went from the back door anyway. The garage itself would receive a promotion, from a house for a car to a shed for the garden. Vegetable beds on the far side of the patio at the end of the driveway-turned-pathway would complete a BBQ-friendly outdoor living room adjoining the soon-to-be-carless garage, all just steps from the back door.

Initially indifferent to the idea of a whole-yard makeover, I came around to seeing the outdoor half of our lot as an integral part of what made our house a home. And now that we had the vision, we needed the plan. So we hired a landscape architect to help flesh out our ideas and draw up a blueprint we could work from. We decided to start with a blank slate. Complete demolition offered certain clarity. If everything was going, we couldn't argue about what to keep. Goodbye to the lawn of foot-stubbing dandelion hillocks, the clump of Japanese knotweed at the end of the driveway, and the Douglas fir sucking up light and space along the back fence.

On top of our glacial till we faced other problems. It's standard building practice to scrape up and cart away topsoil because it's too loose to support a foundation. A century ago, our lot had met this fate, leaving us with 2,500 square feet of pale, rock-filled, dusty dirt—the antithesis of tilth and fertility. I doubt the Osterbergs cared much that by stripping the yard back to bare earth they had rewound geologic time back to when a melting glacier exposed the till. We, however, began to care a great deal as we went about trying to establish a garden in worthless dirt. We couldn't wait centuries to replant and grow an evergreen forest and let nature, once again, turn fallen needles and leaves into rich, fertile soil.

A thriving topsoil industry serves gardeners too impatient to abide nature's pace of making soil. In Anne's job as a river steward, she had run into topsoil-mixing operations. At one place in the watershed where she worked she saw house-sized stockpiles of tree stumps and logging slash next to smaller piles of old doors, bits of drywall, and scrap plywood. A series of rusty conveyor belts leading to large trucks ran past the piles.

Few people know that most bags labeled "topsoil" don't actually contain real soil. Peat moss, ground-up bark, little pieces of pumice, an assortment of fertilizers, and other less than desirable things find their way into those bags. And, if you need truckloads of topsoil, you go to a place like the ones Anne had visited. After what she'd seen, though, we decided to make do with what we had.

This was when Anne launched her crusade. If our yard lacked organic matter, she would add it. She would save our soil.

The second fall after planting the garden, I returned from a trip to find our new patio buried beneath a dark brown pile. It was almost as big as the old VW bug Anne used to own. I set down my suitcase and poked at the pile with my foot, sending small grains tumbling down the side. I was relieved there wasn't anything animal-like about it—that left mineral or vegetable. And there was a familiar aroma emanating from the steaming heap. Laying at my feet was the largest pile of espresso grounds I'd ever seen.

This was just the start. Before Anne finished spreading one kind of organic matter on our beds, she began looking for the next, scavenging whatever she could find. Of course, the neighbor's crispy oak leaves were in the mix, as were the wood chips the arborists continued to deliver. Much like Anne's cooking, improvising new dishes with the ingredients at hand, her mixing, stewing, and brewing of organic matter evolved.

Her new obsession with organic matter worried me a little, especially the way she coveted her loot and confessed to dreaming about it. She assured me, however, that she was no quirkier than other gardeners. I watched her revel in the subdued light and drizzle of quiet fall days, wandering around the garden mesmerized, coffee mug in hand, contemplating where and how she would use her organic stash.

So I wasn't surprised when Anne's organic-matter fetish led to the harder stuff. She was particularly enamored with Fecal Fest, an event that heralded the arrival of "Zoo Doo," composted turds from grazing animals courtesy of Seattle's Woodland Park Zoo. Zoo Doo is almost black in color and can be gooey if we've had a lot of rain. It wasn't free, but it didn't stink. Back when the city used to run the zoo, they were legally prohibited from giving away public property. And since the city owned the animals, they owned what came out of them. A nominal fee to load up on Zoo Doo deterred Seattle's frugal gardeners from mobbing the zoo while still moving the stuff out the door. By the time the zoo changed over to a private nonprofit, Zoo Doo was a much-loved resource among gardening Seattleites.

Soil is curious stuff, part mineral and part organic, a weathered blanket of rotten rock and dead things. Soil life is woven into this fragile, living skin, which constitutes a tiny fraction of the 4,000-mile depth from the ground surface to Earth's core. With a typical thickness of less than three feet, soils vary depending on the bedrock, climate, topographic position, and vegetation. This thin blanket keeps terrestrial landscapes fertile. Soil makes the planet habitable for terrestrial life, integrating life with, of all things, death, to make more life. Plants and animals that die pass into the soil and eventually get reworked into more plants and animals. Think of soil as nature's original recycler, repurposing organic wastes long before we started sorting glass, metal, paper, and plastic.

Anne had her own version of recycling. She stashed her precious piles of organic matter by the side of the garage behind a living fence of interwoven dogwood stems after I insisted she stop using the patio. This became her garden laboratory, where she experimented making different mulches. She shoveled soggy espresso grounds into her wheelbarrow, added oak leaves to soak up the excess water, and threw in larger wood chips. She used the scythe-like blade of her Korean hoe to toss everything together, mixing it like a salad. Season by season, she covered the planting beds with her homemade mulch and watched what happened. If she liked the result, she'd try the same thing again. If not, she kept experimenting. Year-round she layered her organic loot on top of the garden beds. Done early enough in the spring, it meant no watering until July. Done early enough in the fall, it protected plants from freezing.

Anne's mulch recipes were ad hoc, loosely following a composter's rule of thumb. Combine about thirty parts carbon-rich stuff (like wood chips or fallen leaves) with one part nitrogen-rich stuff (like coffee grounds or grass clippings). The exact ratio isn't as important as remembering which materials are carbon-rich and which are nitrogen-rich—and using a lot more of the former.

After a while, adding organic matter to the garden started to feel like painting the Golden Gate Bridge. Just as Anne thought she'd reached the end point, she'd look back at the bed where she started and discover it needed more. We both puzzled over where it was going. Anne jokingly suggested other gardeners were absconding with her precious loot. Then toward the end of the second summer, she complained that the four to five inches of wood chip/oak leaf mulch she layered on top of the soil was rapidly disappearing. The mulch bulked up the soil for a while, then everything deflated like a fallen soufflé. How did all that organic matter break down so fast? More to the point, who and what was breaking it down? And why were they so ravenous? As tons of organic matter disappeared into the silent world beneath our feet, Anne declared that so long as the soil was hungry, she would feed it.

We'd agreed long ago, for the sake of the garden and our marriage, that she is the gardener and I am an engaged onlooker. But, from time to time, she invites me to stick my nose into things. When she complained about disappearing organic matter, I poked around underneath all the mulch she'd been layering on the planting beds. I found bits of the original material, but the rest had vanished like Houdini. Still, our plants were thriving and starting to cover our barren lot in lush greenery. I also noticed how the surface of the soil had changed color, to a brown hue somewhere between milk chocolate and dark chocolate. I figured the darkening of our soil was due to the breakdown of organic matter into humic acids. On average, about half of the carbon in decaying organic matter remains in the soil as nutrient-rich, but decay-resistant, compounds that build fertility like nothing else in nature's store. The other half is lost to the atmosphere during the decay process.

We started with rocky compacted dirt and now we had dark chocolate-brown soil—yes, still with lots of rocks, but not nearly as compacted. Adding organic matter not only enriched the soil, but drew new tenants—mushrooms, bugs, and beetles, and as we later realized, a world of life too small to see.

In addition to feeding our hard, clay dirt with organic matter, Anne watered it, and our new plants, with a home-brewed concoction. Soil soup is an aerobically brewed compost tea she'd learned about at the Northwest Flower and Garden Show. It made sense to her as a biologist and a gardener: culture the desirable microbes found in compost and add them to the soil. But don't do it in a half-assed way; culture a whole lot of microbes, *trillions* of them. Her home-brew package consisted of a six-and-a-half-gallon bucket, a whirling thing that hung on the inside to aerate the water, a gallon of "nutrient solution," and a bag of worm compost (the source of microbes). The whirling thing oxygenated the water during the eight-to-twelve-hour brew period, and the carbohydrates in the nutrient solution fueled microbial growth. She soon devised her own kelp-and-molasses mixture to nourish her home-brewed microbes. Then all she needed was something sturdy and durable to dole out this precious compost tea. A sprayer was the obvious answer. But nothing plastic emblazoned in several languages with "Danger" or "Caution" would do.

She spied what she wanted in one of the many garden catalogs starting to pile up around the house. It was an Old World beauty. An English trombone-style sprayer made of welded brass, wooden handles, and leather washers. I thought it looked like a steampunk pool cue, with a brass nozzle at one end and a three-foot-long pliable hose and filter at the other. She would drop the hose into a bucket of microbial brew and pump the wooden handles back and forth, which drew soil soup up from the bucket and sprayed it out the nozzle in a fine mist. Our trees were small enough at the time that several pumps of the sprayer would leave them dripping compost tea. She'd always save a couple of watering-cans worth of compost tea and use them to water any plants that appeared ailing or those she especially prized.[1]

I thought Anne was delusional to believe that repeatedly dowsing plants with microbes could really make the garden grow better. While I didn't think it would do any real harm, what good, I wondered, could come from spraying the plants with microbial fairy dust and sprinkling it into the soil?

More than I imagined, as it turned out.

At first Anne couldn't find any earthworms in our soil. But, after a year, she discovered fat, liver-colored worms while digging up plants. And I swear that when she showed them to me I caught hints of espresso wafting off their wriggling bodies. A year after that she showed me how some holes, and the roots of some plants, contained weird writhing knots of worms. Around this time, when she was moving mulch aside to dig up a plant, she started noticing big beetles in stout, shiny black armor lumbering out from between tiny twigs and sticks, like tiny bears roused from hibernation. With a clunky gait, they scurried away from her as fast as they could to squiggle their way back down into the mulch. Earwigs, with whom she had a love-hate relationship (mostly hate), dove quickly beneath crumpled leaves and slithered between pebbles when she disturbed the soil.

Mushrooms popped up on the lawn and in the planting beds. And she found webs of fine, white threads stitching together clumps of wood chips she'd put down a year before. Later we learned this stringy stuff was hyphae, the root-like parts of fungi.

As shrubs and flowers matured, insect pollinators arrived. Bees really did bumble around clumps of flowers. In the heat of summer, turquoise and yellow dragonflies showed up, bobbing their way around the garden. The following year, we began dodging plump-bodied spiders that persistently built their webs across pathways and in between trees. On drizzly fall days the tiniest of water droplets caught on their webs, transforming the garden into an eerie setting for Halloween. These ghostly spiderwebs began filling with tiny winged things, which the spiders quickly anesthetized and tidily wrapped up for dinner.

The birds came next. Crows and jays began dropping in, scratching in the mulch atop planting beds with their feet and beaks, as though they had acquired pitchforks and trowels. To them the beds were a smorgasbord filled with tasty morsels ripe for the taking. Flickers and robins started using the lawn as a breakfast buffet, somehow able to sense just where to probe the moist soil with their long pointed beaks.

One of the later arrivals, the smallest bird of all, was the pushiest. Hummingbirds hovered near us if we blocked their access to a favorite flower, furiously beating their wings until we retreated. In the spring these tiny lightning bolts would stage impressive courtship displays. A drab female perched on a needle-thin twig attentively watched a colorful jewel of a male shoot skyward and acrobatically dive-bomb the yard at a breakneck pace, pulling up at the last second before impact.

As the garden matured, larger animals started showing up too. A rocket-fast Cooper's hawk began cruising through, looking for unwary smaller birds to nab. Bandit-masked raccoons staked their claim year-round. Looking back, I realize that making a garden had given us a ringside seat to the grand march of life. Flying, crawling, walking, and scampering they came, and we witnessed it all.

Our plants were thriving too. Anne's soil soup, mulch, and compost concoctions had delivered the desired effects. The plant diseases and pests that other gardeners battled either never cropped up or, if they did, never took hold. Our shrubs had huge leaves and the trees were as happy as could be. Anne even snuck in a couple of climbing roses and none of the usual Northwest rose ailments—like mildew or blackspot—afflicted them so long as they got their monthly dose of soil soup.

About three years after we started building the garden, organizers of a north Seattle garden tour asked if they could include ours in the year's event. Judging by the interest from complimentary gardeners, we knew we'd done something right—we just weren't sure how it worked.

Curious neighbors wanted to know Anne's garden secrets and peppered her with questions. What made the trees grow so large so fast? A coral bark maple and a set of Persian ironwoods, all had trunks little bigger than golf club shafts when first planted. In less than a decade, their size approached the girth of an elephant's lower leg. Neighborhood gardeners wondered why the same plant in their yard, a block away, was ailing. Was it the amount of sunlight in our yard? The amount of watering we did, or special fertilizers we used? Not a soul asked us about the soil. It was never mentioned: nature's greatest wallflower. Anne gave impromptu tours of the garden to interested passersby and pushed her soil soup and mulch recipes. But everyone kept their gaze aboveground, asking her about the plants they could see and touch.

Five years after we'd planted the garden, the afternoon light was just at the right angle for me to notice that the lawn was creeping onto the patio. Not just the grass, but also the soil, the ground itself. I remembered that the soil at the patio's edge had started out flush with the pavers. Now it towered a quarter inch above the patio, forming a miniature cliff held in place by a web of fine roots. And the exposed soil was deep brown, no longer the khaki color I remembered digging into when we first put in the garden. The earth had been changing right before my eyes, and right under my nose—just too slowly to notice day-to-day.

Sitting at our book-laden dining room table, I watched Anne through the old wavy-glass windows. She moved in and out of my view, puttering in our growing garden. She'd splurged and bought a brand-new wheelbarrow and then, in a fit of inspired creativity, painted swirling flames on it. Overfilled with her scavenged organic matter, it teetered and listed as she wheeled it from planting bed to planting bed. The significance of what she had been doing these past few years dawned on me in the context of research I was doing at the time on soil loss. Anne was demonstrating the solution to an age-old problem, one wheelbarrow load at a time. She was building new soil far faster than nature could.

Scattered through history, a few societies bucked the historical trend of soil loss, among them Amazonian Indians, Asian peasant farmers, and nineteenth-century Parisian city gardeners. The common element? Precisely what Anne had been doing to our yard: returning organic matter to the land. My gardener-wife was bucking the trend that had destroyed societies around the world. Of course, Anne was too busy in the garden to think about how she was making soil faster than nature. Yet as the soil got darker, the digging got easier, and we began to see connections between our disappearing organic matter and the explosion of life in our new garden. Seeing tons of organic matter disappear into a silent ravenous world beneath our feet, it slowly dawned on me that all the compost, mulch, and soil soup Anne had been adding to the earth was the solution for how to make fertile soil—and how to do it remarkably fast. Was it possible that thinking like a gardener could help humanity build fertile soil on a far larger scale, rather than use it up?

It took a little over half a decade to restore life to our yard. This may seem like forever in today's ultraconnected, high-speed world. But to a geologist this is faster than the blink of an eye. As Anne and I watched our yard evolve from a biological void into a place thriving with life, we were most struck by how, of all things, dead stuff—organic matter—spawned a web of new life. Mulch, compost, and wood chips fed the life of the soil. And that fed plants, animals, and eventually even us. We hadn't anticipated how the revitalized garden would draw animals, but one day we were schooled in how life below ground shapes life above ground.

Several years after we planted the garden, Anne and I went outside to drink in one of Seattle's flamboyant summer sunsets when she noticed a dark shape slowly approach and soar over our heads to a neighbor's sixty-foot-tall Douglas fir tree. Without missing a wing beat, a bald eagle plucked a baby crow from its nest in the spire of the tree and flew off, a clump of downy dark feathers dangling from its talons. Seconds later, a black funnel cloud of frantic crows descended on the nest tree cawing an avian dirge. For half an hour, we stood transfixed in the middle of the garden mulling over the eagle's visit as a murder of crows swirled and screeched around us. The cycle of life that began with us reviving the soil culminated in an eagle hunting the baby crow whose parents ate the worms that grew in our soil. The garden became a microcosm of the cycle of renewal and decay that turns the global wheel of life.

Gardeners fussing over the plants on top of the soil might look beneath their feet—to the microbes and invertebrates that call soil their home. People think soil is static and inert, like the rocks from which it comes. But watching our soil change opened our eyes and then our minds.

Just as in the evolution of life on Earth, microbes and soil life set the stage for all that follows. The order in which life came to the garden mirrored the order in which life evolved on Earth—from microbes and fungi, to worms, spiders, beetles, and birds. Microbes long ago showed the way for life to leave the sea; they also were the first to colonize the new continent of our embryonic garden. What unfolded beneath our feet and in front of our eyes showed us a fundamental truth about terrestrial ecosystems—microbial life is the foundation supporting them all. Like an iceberg, the natural world visible to us above ground is kept afloat by what's beneath the surface.

Paying attention to the natural world is an ancient human habit. Knowing about nature used to matter a great deal before we began to live in cities, drive around in cars, and sit in front of screens all day. Once upon a time, we spent our lives figuring out what to eat, where it lived, and how it grew. Before the dawn of agriculture and civilization, one had to know the difference between plants. Mix up the leaf shape of a delectable plant with that of a poisonous one and, if you lived, you would tell your story to others, vowing never to repeat the mistake.

Back then, you studied animals that traveled as a herd and noticed the habits of fish in a river. You could return to those places and catch dinner, eat it on the spot, or carry it back to the clan to put over the fire. Paying close attention to the natural world meant the difference between having plenty or going hungry; it meant the difference between life and death.

But today we have grocery stores and restaurants and quite a bit less nature to see and wander through. Knowing one plant from another, or where to find the best fishing hole, is no longer a requirement for a long life. Nonetheless, the instinct to connect with nature resides in us. Caring for a little patch of nature—a garden—engages this instinct in ways that few other things do. Cultivating plants (and animals) for food, comfort, aesthetics, and pleasure is as old as civilization itself. But our natural inclination to focus on the nature we can see has led us to overlook the importance of the half we can't. Even the early naturalists neglected microbes during the heyday of taxonomy. For how could something so small do anything of much importance?

The hidden half of nature is our name for the abundant and little-known life forms too small to see that live beneath our feet and, as we'll see, within us too. The process of solving the mystery of where all of Anne's organic matter went unleashed our curiosity. What else were microbes capable of? More than we ever imagined.

2

THINKING SMALL

Humanity paid little attention to microbes for so long for a reason so simple a toddler would understand. We can't see them. Out of sight, out of mind. Literally.

Indeed, their invisibility defines them. Any creature too small to be seen with the naked eye—anything smaller than a tenth of a millimeter—is considered a microbe. Bacteria are the most studied type of microbe, and one of the most well known is *Escherichia coli*, *E. coli* for short. At two microns long—that's two thousandths of a millimeter—it would take almost 40,000 of them laid end to end to wrap around your thumb, and 100 of them to span the period at the end of this sentence. If this is hard to imagine, consider that were a bacterium the size of a pitcher's mound, you would be about the size of California.

Despite their diminutive size, microbes are the most abundant, widely distributed, and successful organisms on Earth. More than 99 percent of all species whose bones are preserved in the fossil record are extinct, having failed the test of time. Yet microbial life has survived since the earliest days of life, more than 3.6 billion years ago. This pencils out to at least 80 trillion generations, given their short life span.

In all, there are an estimated 10^{30} microbes on Earth. That's a nonillion, a 1 followed by 30 zeroes—1,000,000,000,000,000,000,000,000,000,000. Although each microbe is too small to be seen, link them all together in a chain and they would stretch 100 million light-years—well beyond the farthest visible star in the night sky. Microbes on our planet outnumber the stars in the known universe more than a million times over. More bacteria live in a handful of rich fertile soil than the number of people who live in Africa, China, and India combined. And all together, microbes are estimated to make up half the weight of life on Earth.

Microbes are as diverse as they are numerous and fall into five main types—archaea, bacteria, fungi, protists, and viruses. Although distinct species are a slippery concept when it comes to microbes, biologists estimate that there are millions to hundreds of millions of them in the microbial world.

Archaea (r-KEY-uh) are the most ancient. They were once thought to be bacteria, but the chemical composition and structure of their cell walls are utterly different from those of regular old bacteria. Some types of fungi are microscopic, like yeasts, while other types, like the above-ground part of mushrooms, are easily seen.[1] Protists include amoebas, diatoms, and a raft of other strange-looking one-celled organisms. Fungi and protists house their DNA in a nucleus; archaea and bacteria do not.

Viruses confound us. They are not made of cells, and aren't alive, even though they do life-like things. Some scientists think viruses are microbes, others do not. But viruses do take advantage of life. A type of virus called a bacteriophage (because it infects only bacteria) can dock on a bacterium's surface like a tiny spaceship and inject its genetic payload into the cell. This kicks off the viral reproductive cycle, which tricks the host bacterium into making a lot of copies of the virus—at its own expense.

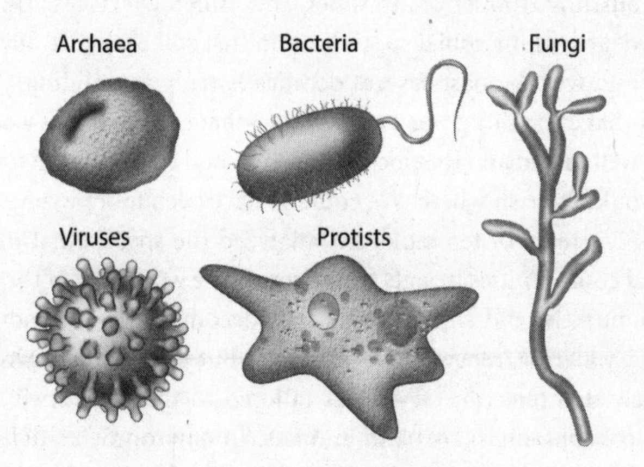

Portraits. The major groups of microbes.

Archaea are not currently thought to cause diseases in plants or animals (including people), while fungi are major agents of plant diseases and also can make people sick. Bacteria, protists, and viruses are the dominant agents of animal and human diseases and wreak havoc on plants too.

Particular antibiotics are effective against different types of microbes due to differences in their cell structure.[2] Drugs effective against bacteria do not generally harm archaea. Antibacterial agents don't affect fungi or viruses; and antifungal drugs don't work on bacteria, though they are sometimes hard on our systems.

Try as we might to push the thought of microbes out of our minds, they are pretty much everywhere—on every natural surface, in every drop of water, and on every grain of sand. Over the past few decades, scientists have consistently found microbes wherever they've looked. With recently formalized projects like the Earth Microbiome Project, there is a major scientific effort under way to understand and map the global microbiome—the secret basement in the house of life, a place far wilder, more widespread, and more influential than previously imagined.

As humanity explores the world of microbes, we're finding that we know far less about their ecosystems than we do about those we can easily see—Earth's oceans, forests, rivers, and deserts. Alive with microbes, soil is a bustling frontier of air, water, and mineral surfaces. The diversity and degree of microbial specialization that soil ecologists have been discovering over the past several decades is truly astounding. The old idea was that everything was everywhere—that the microbial world was globally well mixed and that local environmental conditions determined which would flourish where. We couldn't have been more wrong.

In 2012, a team of ten biologists analyzed the species that make up microbial communities in soils from around the world. All of the microbial communities did similar tasks, like decomposing organic matter, purifying water, or renewing soil fertility. But each community had a completely different mix of species tailored to the local environment and the cohabitating plants and animals. In environments dense with life, like tropical forests, the biologists found a far greater diversity of microbes involved in the endless task of decomposing organic matter. But, because there is so much less organic matter in colder, drier environments, like deserts and polar regions, fewer species perform that task. And just like in the above-ground ecology we know, loss of a single species below ground can ripple through a whole community, causing dramatic changes. Unlike the macroscopic world, however, we have little grasp of who the key players are in microbial ecology—let alone how they may work with or against one another in different circumstances.

Recent advances in geobiology, a burgeoning subfield of geology, show us that microbes are also biology's best globetrotters. Among all life forms, they occupy the highest highs and the lowest lows. Bacteria roam Earth's lofty ceiling of the upper atmosphere, growing in cloud droplets. And archaea live near boiling volcanic vents on the deep seafloor.

Microbes occur nearly everywhere because they are so remarkably adaptable. They survive where nothing else can, using an incredibly diverse array of things as food. Some archaea known as extremophiles can withstand the worst heat, cold, and dryness on Earth. Geologists exploring the deep seafloor were surprised to find communities of microbes living on and around "black smokers," natural chimneys rising twenty to thirty feet above the seabed, where high pressure keeps water liquid despite temperatures reaching 750°F. Remarkably, archaea have also been found in frozen lakes trapped half a mile beneath Antarctic ice.

Scientists have even found microbial communities in the Atacama Desert in Chile, a place with no rain, no streams, and no lakes. This environment is where NASA field-tested the Mars rovers. In most places on Earth, no water means no life. Yet, in 2005, bacteria were discovered living below the surface of a salty, dried-up ancient lakebed in the Atacama. How did they survive? Occasionally, enough dew from humidity on cold nights would get sucked into the salt to revive dormant microbes in the crystal matrix, raising the intriguing question of whether microbes could potentially live in salt deposits on Mars.

Recently, some scientists went a step further in their speculations. They suggested that microbial life might have arisen on Mars and then traveled to Earth on a meteorite. Could a microbe really survive the exposure to radiation on the trip from Mars without a spaceship? We already know of microbes that have what it takes.

Deinococcus radiodurans is unbelievably resistant to radioactivity—as well as extreme heat, cold, and exposure to acid. In the 1950s, researchers applied lethal levels of radiation to a can of meat and then opened it up. Inside, they found an unscathed colony of *D. radiodurans*. This resilient bacterium can tolerate a thousand times the radiation necessary to kill a person, a trait that allows them to thrive in the deadly cooling ponds of nuclear power plants. Scientists hope one day to genetically modify *D. radiodurans* and unleash them to eat, and thereby clean up, nuclear waste.

PRIMORDIAL HOLDOUTS

Many of today's extremophiles are the descendants of ancient archaea, dating from the early days of life when Earth was a superhot planet with little oxygen. Some of these descendants still cannot tolerate oxygen. We call them anaerobes, and they harvest energy from a range of compounds, such as hydrogen sulfide, methane, and ammonia.

The vast genetic repertoire of microbes allows them to use virtually any naturally occurring element or compound as an energy source. The enduring resilience of microbes is all the more remarkable considering the dramatic changes to their habitat over the past several billion years. When photosynthetic bacteria (particularly cyanobacteria) evolved, they started pumping their metabolic waste—oxygen—into the atmosphere. The oxygen readily reacted with a wide range of elements, chief among them iron. Over time, bacterial activity catalyzed the precipitation of a huge amount of iron oxide out of primordial seas, creating the distinctive banded-iron formations that make up some of the world's oldest rocks. As a result, oxygen levels in the atmosphere rose and fell repeatedly. But between 2.5 billion and 2.3 billion years ago, the growing cyanobacteria population depleted the supply of readily oxidizable minerals, and oxygen began to accumulate in the atmosphere. A billion and a half years later atmospheric oxygen reached levels high enough to support animal life. The persistence of today's oxygen-rich atmosphere still depends on microbial mediation of the global oxygen cycle.

While the success of photosynthetic bacteria paved the way for us oxygen-breathers, it poisoned the world's surface for their anaerobic brethren, the archaea. That's why most archaea now hunker down in the planet's crust of fractured rock. But some live sheltered inside plants and animals too. By some estimates, archaea still account for as much as a fifth of all the biomass on the planet. We just never see them.

A surprising find in the mud beneath the seafloor off the Peruvian coast illustrates the ancient roots of the modern biosphere. A core drilled five hundred feet down revealed a vibrant microbial ecosystem of anaerobic archaea, bacteria, and fungi that use sulfate and nitrate as energy sources in their sunless environment. More surprising is that these microbial communities live in 5-million-year-old sediments. They have been isolated from life on the surface for longer than humanity has existed.

Go down even deeper, into the basaltic rock beneath the deep-sea mud, and you'll also find microbial communities. In 2010, a team of earth scientists reported evidence of abundant microbial life in fluids seeping from natural vents off the coast of Washington and British Columbia. Residing far too deep for photosynthesis, these microbes living in Earth's watery basement metabolize sulfates dissolved in the water that circulates through crustal rock. The surprising discovery that the basalt beneath the seafloor could be chock-full of microbial inhabitants means that Earth is literally riddled with life, and most likely has been for billions of years. We share the planet with an entire world of subterranean life that we are just finding out about.

ANCIENT TRADERS

How did microbes get all over the place? It helped that as other life forms evolved, bacteria hitchhiked around the world on plants and animals. They also blew in on the wind.

Every day, hordes of soil bacteria, in dust kicked up from arid regions in Asia, land in North America, carried on transpacific winds. In 2011, scientists trapped some of these microbes at the Mt. Bachelor Observatory more than 9,000 feet above sea level in the Oregon Cascades. They reported finding thousands of species from both marine and terrestrial sources.

Why are microbes so diverse and seemingly able to occupy any and every niche? Chief among the reasons are lightning-fast reproduction, and the way microbes acquire their genes. One of the most surprising things scientists discovered in the last half of the twentieth century is that bacteria, archaea, and viruses swap genetic material like we swap stories. And not just among themselves, but well beyond their kind. Microbes pass genes to protists, to insects, to plants, and to animals. They do not abide by the rules we do. Their sexless genetic promiscuity has the least titillating name imaginable—horizontal gene transfer. This shocking behavior violates all the rules in the Darwinian playbook.

Bacteria are as opportunistic as dogs, scarfing up DNA straight off the floor of their environment. One recent study even discovered that bacteria had incorporated DNA from a 43,000-year-old wooly mammoth bone into their own genome. This sets up an entirely different evolutionary game for microbes than for larger organisms, which must bother with courtship and the clumsy business of sex to pass on their genes. The ability to casually trade genes and to vacuum up DNA from their surroundings, including the genetic scrap heap of dead things, lets microbes quickly adapt to new conditions.

With no hard parts to fossilize, early microbial life disappeared rapidly upon death. Durable shells and skeletons that could get preserved in the rock record only evolved around 541 million years ago. Nonetheless, evidence for microbial life extends back to at least 3.4 billion years ago. Such is the case for bacteria that left ghostly traces of their existence as imprints in South African rocks nearly that old. Indirect geochemical evidence even points to the presence of life on Earth before 3.8 billion years ago.

Microbes have never been particularly solitary. Most live as colonies in communities of multiple species, a far cry from how they are usually studied in single-species laboratory cultures. Some stick together, literally, coating surfaces with resilient, tough biofilms. But a biofilm is more than a clump of cohabitating bacteria. The glue-like matrix that binds them together comes from a mix of proteins and long chains of complex sugars called polysaccharides that the bacteria themselves secrete. These microbial cities grow everywhere moisture clings to a surface. Biofilms even grow in our bodies. A common example is the plaque that coats our teeth. The recent discovery of gum-dwelling bacteria turning up in biofilms on the walls of arteries suggests that oral bacteria might play a role in heart disease. But don't go thinking biofilms are all bad. As we'll see, in some settings biofilms of beneficial bacteria help keep detrimental bacteria from gaining a foothold.

Biofilms were even instrumental in launching life. The earliest-known fossils you can see with your own eyes are stromatolites, finely layered rocks that record the growth of bacterial mats—biofilms—in shallow marine waters. In 2008, analyses of organic globules preserved in 2.7-billion-year-old stromatolites supported their microbial origin. The stromatolites formed when thin layers of sediment settled and subsequently became trapped among the filaments and mats made of bacterial colonies. While the individual bacteria involved were not preserved, they left behind evidence of their biofilm structures, the dominant form of direct evidence for early life on Earth. In 1956, living stromatolites were discovered in Shark Bay, Australia, making them a rare example of a life form alive today that was first discovered in the fossil record.

The global effects of microbial life, especially photosynthetic bacteria, are hard to overstate. In addition to creating Earth's oxygen-rich atmosphere cyanobacteria living near the surface of the oceans help regulate atmospheric CO_2 for the entire planet. They draw carbon out of the atmosphere through photosynthesis. Should a substantial proportion of these bacterial photosynthesizers suddenly die atmospheric CO_2 levels would rapidly rise. Just such a rise occurred at the end of the last glacial period, too quick to be accounted for by geophysical or geochemical processes. Could this mean that a sudden decrease in microbial photosynthesis helped usher in the postglacial climate in which humanity subsequently rose to global prominence? Perhaps.

And we can credit (or curse) microbes for the supplies of fossil fuels that we have been burning so fast that it is increasing atmospheric CO_2 to levels the planet has not seen in millions of years. Petroleum and natural gas are composed of either the remains of microbes or microbially decomposed organic matter heated up and cooked at high pressure deep in Earth's interior. We can also thank microbes living closer to home for the aroma of farts, which they produce courtesy of the anaerobic breakdown of proteins.

More important, microbes capture the atmospheric nitrogen essential for making amino acids needed to sustain life; they run Earth's nitrogen cycle, nature's way of keeping soil fertile. The nitrogen content of rocks varies greatly, from trace levels in granites to biologically useful levels in some sedimentary rocks. Throughout geologic time almost all of the nitrogen in organic matter—from simple proteins to DNA, the molecule that rules us all—entered the biosphere through microbes.

You can catch a glimpse of this process if you pull up a pea plant, or one of the other 10,000 species of legumes. Growing on the roots you will find bulbous nodules. Cut one open and you'll see a blood-red color. This is a telltale sign of the chemical reaction that special bacteria oversee—the conversion of nitrogen from the atmosphere into a soluble form that plants can use. Without the continuous work of nitrogen-fixing bacteria to replenish forms of nitrogen that living things need, the grand enterprise of life would rapidly grind to a halt.

Microbes themselves have benefited from sustaining the global cycles that run our planet's ecosystems. When the first intrepid amphibians left the seas to explore land, microbes came along for the ride. Evolution of the rigid, water-holding structures we know as plants also helped microbes expand into terrestrial environments. Try thinking of plants and animals like spaceships for adventurous microbes setting off to explore new worlds.

But microbes are more like a crew than like stowaways. From roots to fruits, plants are coated with them. Long ago, microbes colonized the interiors of animals and made themselves indispensable in the digestive process, cementing partnerships with a wide range of species, from aphids to cows to clams.

It's been hard to appreciate the diversity of microbes and the great many things they do because our notions of diversity come from the part of nature we see every day—the visible realm. We know plants and animals in terms of shape, size, and color, traits that catch our eye or matter to us. But the diversity of microbes is expressed in a completely different way. It lies not in what they look like, but in the tremendous variety of molecules and compounds that they can use and make.

We can't eat rocks, yet our bodies are made of nutrients that come from rocks. Microbes play key roles in breaking down and extracting elements from rocks and getting them into biological circulation. And consider that animals, including nearly all insects, cannot actually digest plant matter made of cellulose, a very stable and hard-to-breakdown molecule. Although cellulose is the most readily available food (and energy) source in the world, animals delegate the difficult task of decomposing it to the microbes living in their gut.

COW POWER

Cows, as we all know, eat grass. But without the microbes that make the enzymes to break down cellulose, cows would starve no matter how much they ate. Communities of specialized microbes live in the rumen (the first compartment of a cow's four-chambered stomach), where, day in and day out, they do the hard work of breaking down cellulose. So cows don't simply eat grass; they chew it up and feed it to their internal microbiota, who break it down and offer up nutrients in return. In other words, cows don't chew their cud because they are content. They do it to survive. They must grind grass into particles fine enough for microbes to digest the cellulose. Cows graze quickly and then regurgitate small amounts of grass to grind into finer pieces, ruminating for up to ten hours a day. They are four-legged power plants that turn grass into energy.

The physical apparatus that allows cows to pull off this feat of internal farming is their elaborate four-compartment stomach. The rumen alone contains about a quadrillion microbes. They release the enzyme cellulase into the rumen, which breaks cellulose down into digestible sugars. The fine-ground cud increases the surface area on which the cellulase acts. The cow's second stomach, the reticulum, acts as a mixing chamber that extends the microbial fermentation that began in the rumen.

If this sounds like a too-good-to-be-true deal for a cow, that's because it is. Cows don't directly benefit from the sugars their microbial crew produces. The microbes consume the sugars themselves, producing short-chain fatty acids like acetate, propionate, and butyrate that the cow then absorbs. In other words, a cow lives off the waste products of the microbes it hosts. As we'll see, they're not the only mammal that does.

But the cow gets the last laugh. It also eats the microbes. The third part of the bovine digestive tract, the omasum, is a muscular vessel that contracts, pulling the most digested components through a small orifice to absorb water out of the material before passing it along to the fourth chamber, the abomasum. This final stop is where the cow digests the cellulose-eating microbes that grew in the rumen. These microbes are the cow's major source of protein.

This comes at a price, however. The methanogenic archaeans that cows rely on to help digest their diet—whether grass or grain—produce a lot of gas. Archaea-produced methane can build up and bloat a cow. Unless, of course the cow eructates (belches) out the gas. A typical cow can belch out more than twenty-five gallons of methane a day. Add it all up and livestock account for a third of U.S. methane emissions—more than oil and gas. And while some vilify them as huge emitters of greenhouse gases, cows themselves don't actually produce the methane that farm kids delight in putting a match to.

Cows eat grass to feed their internal microbial fermentation tank. And in exchange they live off the products of microbial fermentation—and the microbes themselves. We can't eat grass, but thanks to the internal microbial garden in cows, we get to drink milk and eat cheese and roast beef.[3]

The bottom line is that anywhere there is life there are microbes—from every surface in your house to the most extreme environments on the planet to the four-compartment stomach of a cow. Although microbes started us down the road to complex life, it took until the waning days of the twentieth century to realize this. For a long time after they were first discovered, microbes were considered little more than entertaining distractions.

3

LOOKING INTO LIFE

Long before agriculture transformed civilizations around the globe—and our relationship with the natural world—people lived immersed in the nature they could see and touch. It's no accident that we have the brain capacity to name and remember more than five hundred kinds of animate things. In forests, meadows, and thickets, our forerunners organized useful plants and fungi into the edible, the medicinal, and the poisonous. Without farm manuals and garden books, tractors or trowels, the first people to add plant cultivation to their food-gathering strategy must have been truly remarkable. They thought to nestle seeds into the ground and coax meals from the earth. Since then, many, like Anne, fell for the greenery around them. But few swooned over what remained unseen.

NAMES FOR NATURE

One man, irresistibly drawn to all types of greenery, helped shaped our view of nature. Born in 1707 in the Swedish countryside, he grew up exploring and wandering the natural environments of northern Europe. As a young man, he studied medicine and developed a keen interest in botany, as plants, after all, served as the pharmacy of his day. Later in life, he developed an uncanny ability to order plant life into distinct groups based on their features. When presented with mystery plants from faraway lands, he seemed to intuitively know the group to which each belonged. He classified them based on the features he could see—the form of a leaf margin (serrated or smooth), the number of stamens in a flower (five or fifty), the girth of a seed (apple- or avocado-sized). This remarkable person was Carl Linnaeus.

With ambition to match his skills of perception, Linnaeus proposed a new way of naming and classifying all of life that laid the groundwork for the field of taxonomy. Every plant, whether flowering or thorny, puny or giant, and every animal—trotting, plodding, or soaring—would have a name and a place in his System of Nature (*Systema Naturae*). In what began as a slim pamphlet and grew into a masterful tome, Linnaeus laid out his idea for how to further classify the two most basic categories of life—plant and animal.

Linnaeus's system for naming plants and animals, known as binomial nomenclature, is on display at any plant nursery you walk into. Every plant has its own one-of-a-kind, two-part Latin name—like *Acer palmatum* for Japanese maple and *Vaccinium parvifolium* for red huckleberry. Visit a zoo and you'll likely see *Panthera leo*, a lion, and its favorite snack, *Aepyceros melampus*, an impala. The first part of the name specifies the genus, common to everything closely related. The second part is the species name. Together, the unique taxonomic label of genus and species defines every organism. Naming plants and animals frames the way people look at, talk about, and understand nature.

Linnaeus knew nothing of genetics, predating Gregor Mendel by a century. Nor had he heard of evolution, as Charles Darwin had not yet been born. And of course he couldn't have dreamed of modern-day gene sequencing or DNA analysis. Linnaeus had to rely on his eyes to study the natural diversity of the world around him. His method of categorizing plants and animals worked so well that it is still used today. In laying the groundwork for how to name and categorize all of life, Linnaeus influenced the thinking and work of every biologist and naturalist who came after him. For this, he is considered the father of taxonomy.

Children of course do not know about taxonomy. But somehow they seem to know what makes a plant a plant and an animal an animal almost from the time they start looking beyond their parents' faces. Before they can talk, most kids can readily tell a cat from a dog and a fern from a tree. Toddlers point and shriek with aplomb at the life they see around them. Parents chock it up to youthful exuberance, but it runs deeper than that.

Our innate desire to categorize the nature we can see is a fundamental part of who we are. We rely on our eyes more than any other sense to tell us who is who, who is related to whom, and how they should be grouped. Mostly, we group plants and animals based simply on what they look like. It's one of the first things you learn to do no matter what culture you live in. And what we can't touch or see doesn't count for much.

For over two hundred years, the Linnaean system of perceiving, naming, and classifying life worked perfectly well—until recently. Observing and examining whole specimens, whether alive or dead, and using a ruler to measure a tail, antennae, or petal falls short when it comes to microbes. Woefully short, as it turns out. Even the idea of fixed species doesn't really apply in the microbial world.

Microbes stumped Linnaeus. Discovered just a few decades before his birth, they remained a mystery. Were they plants or animals? In Linnaeus's view, there was no point in classifying the little buggers. They were far too hard to see and all too much alike. You couldn't get your arms around them, much less press them between herbarium sheets or pin them onto a specimen board. In later editions of the *Systema Naturae*, Linnaeus lumped all microbes into a group commonly called infusoria, assigning bacteria to the genus *Chaos*. It would take several centuries, and the discovery of DNA and subsequent technological innovations in DNA sequencing, before the intimate ties of the microbial world to the everyday world we know would start to come into focus.

Your ability to tell fish from frogs, and sunflowers from lilies, is of little use when it comes to creatures invisible to the naked eye. You will never see toddlers chortling over a bacterium the way they do over a kitten or a bird. Microbes are invisible. And therein lies the problem. Our yearning to touch, see, and classify the natural world around us falls apart when it comes to life too small to see. But that began to change once a Dutch draper opened a window onto the smallest creatures on Earth.

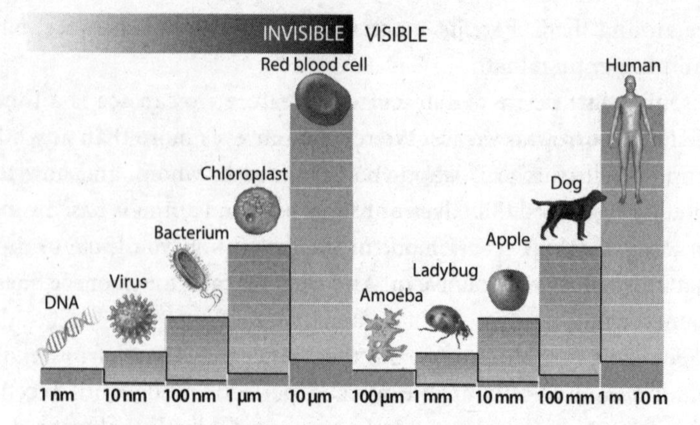

Sizing Up Life. The world of life spans nanometers to tens of meters, from the not-really-alive viruses to the nature we see in our everyday lives.

WEE BEASTIES

When Antony van Leeuwenhoek (LAY-vin-hook) discovered microbes, he considered them nothing more than amusing curiosities. The son of a prosperous basket maker, Leeuwenhoek was born in 1632 in Delft, Holland. At age sixteen he left the boarding school his mother sent him to after his father died and secured an apprenticeship as a bookkeeper with an Amsterdam cloth merchant. In the merchant's shop he saw his first microscope, a rudimentary device consisting of a simple lens on a small stand used to inspect the quality of linen and draperies. Six years later, in 1654, he returned to Delft to set up his own drapery store, soon marrying and settling into a comfortable life. Unable to read Latin, the lingua franca of scholars, he knew nothing of the dawning world of science.

As the years rolled by, Leeuwenhoek developed an intense interest in grinding better glass lenses to inspect his merchandise. He became enthralled with the microscopic world after seeing stunningly detailed images of fleas and other tiny creatures in a copy of the first scientific bestseller—Robert Hooke's vividly illustrated *Micrographia,* published in 1665. Wonder blossomed into obsession, and Leeuwenhoek began visiting spectacle makers and alchemists to learn the secrets of lens crafting and metalwork. A few years later, in 1671, he mounted one of his finely crafted lenses on a stand-like contraption and made the most powerful microscope of his day. This inspired tinkering led Leeuwenhoek to engaging distractions and unimaginable discoveries.

His microscopes were just three to four inches long, but far more powerful than anything available to merchants and scholars. They made elk hair look like logs and flakes of his own skin look like scales. He even dissected the head of a fly, marveling at the complex detail of its minuscule brain stuck on the head of a pin. Of course, his bourgeois neighbors thought him crazy. But Leeuwenhoek didn't care; he was absorbed in exploring miniature worlds. And he did not advertise his astounding discoveries, either—they were for his entertainment alone.

Eventually, Leeuwenhoek let an acquaintance who was a correspondent with London's Royal Society peer through his microscopes. The experience astounded the worldly, well-educated visitor. The best hand lenses of the day could magnify things severalfold. Yet, here in a merchant's parlor, you could inspect the legs of a louse and other marvels magnified to more than two hundred times their actual size. Word of these amazing microscopes soon reached the Royal Society.

Then, on a windy September day in 1674, Leeuwenhoek collected some water from a pond near Delft in a glass vial, curious as to the cause of milky green clouds floating in it. When he examined the water through one of his microscopes, he saw tiny green filaments the width of a human hair, made of globules joined end to end. Tiny creatures danced around in this microscopic forest. A year later, in September 1675, Leeuwenhoek examined a drop of rainwater from his garden that had sat for four days in a blue tub. After peering through the lens, he called for his daughter, Maria, to come and see the ocean of life in this raindrop sea. Wee creatures darted about beneath the lens. These little animals lacked heads and tails. Most looked like blobs or sticks. Some resembled corkscrews. The smallest were barely visible, their shapes murky.

Where did the wee beasties come from? Did they fall from the sky with the rain? Did they live on the ground where the raindrop fell? Or had God brewed them up right in his garden?

The following spring, in April 1676, Leeuwenhoek placed a third of an ounce of pepper grains in water, curious about whether their spiciness was truly due to tiny barbs as commonly believed. After letting the pepper steep for three weeks, he didn't spy the tiny hooks he expected to see. Instead he found a diverse community of what he called "animalcules" darting about beneath his microscope.

Then, during a heavy rainfall on May 26, 1676, Leeuwenhoek grabbed a wineglass, thoroughly cleaned and dried it out, and held it under the downspout from the eaves of his roof. Placing a drop of the rainwater from the glass beneath a microscope revealed scores of little beasts swimming about. He cleaned out a blue-glazed porcelain dish and left it out in the rain. When he placed a drop of just-fallen rainwater beneath his microscope, nothing wriggled past. It seemed the tiny animals did not arrive with the rain. Or did they? Four days later, the same water was alive with these mysterious creatures.

Excited by these discoveries, Leeuwenhoek wrote to the Royal Society describing how a single drop of water could host a million invisible "beasties." The learned men of London scoffed. This was absurd! How could a draper from Delft discover more animals hiding in a drop of water than there were Dutchmen in all the world? Everyone knew that the smallest of God's creatures was the tiny, almost-invisible cheese mite.

Leeuwenhoek received a skeptical reply asking him to detail both how he made his microscopes and his methods of observation. He responded with calculations about estimating the size of these tiny creatures and an offer to send affidavits from prominent citizens of Delft who would swear to having seen them with their own eyes. But, secretive to the point of paranoia, he would not reveal how he made his microscopes.

Intrigued by these claims, outrageous as they sounded, the Royal Society commissioned Robert Hooke to build the very best microscope possible and test Leeuwenhoek's wild assertions. On November 15, 1677, Hooke brought his new high-powered microscope to a meeting of the society, where the astonished membership took turns peering at swarms of tiny creatures swimming in a drop of water. Duly impressed, the Royal Society subsequently elected the Dutchman a Fellow.

Still, Leeuwenhoek would not send them a microscope. His inventions were not for sale.

Leeuwenhoek looked for life in other places too. He scraped the plaque from between his teeth and placed it in a drop of pure rainwater. It swarmed with tiny creatures. Some leapt around like fish, others tumbled in endless somersaults. Some moved sluggishly, others darted frantically about. Even his mouth was alive with tiny rods and animated corkscrews.

Later, he looked again at plaque from his teeth, this time collected after drinking hot coffee. Only a few creatures moved around feebly. But he could make out scores of dead ones. It seemed heat killed the little creatures.

On a roll, Leeuwenhoek kept looking and found microbes in the strangest of places—drinking water, the intestines of horses and frogs, and even his own diarrhea. Along the way, he became something of a celebrity, discovering blood cells and human sperm. He frequently entertained prominent visitors curious to look through his famous microscopes, including Tsar Peter the Great of Russia and Queen Mary II of England, both of whom traveled to Delft to peer through his magic lenses.

Leeuwenhoek remained obsessed with microbes throughout his life and on his deathbed. In 1723, at the age of ninety-one, with his daughter Maria by his side, he mumbled instructions for a final letter to the Royal Society. To the end, Leeuwenhoek largely considered his wee beasties a ceaseless source of wonder irrelevant to our everyday world.

FERMENTING GENIUS

More than a century later, scientists began to realize how much microbes influence our everyday lives. In October of 1831, an eight-year-old boy from a mountain village in eastern France asked his father why the bite of a rabid dog could kill a healthy man. The elder Pasteur, a tanner who served as a sergeant in Napoleon's army, had no idea. But the question haunted young Louis, an inquisitive child destined to change the way we see nature's hidden half.

A solid, though uninspired, student, Pasteur gained admission to the École Normale Supérieure, a prestigious teacher's college in Paris. There he fell under the spell of Jean-Baptiste-André Dumas, the father of organic chemistry. At the time, microbes were receiving renewed attention after reports that live yeast were essential for converting a brew of hops and barley into beer, and that meat could remain fresh for months if heated in a sealed glass jar. It was starting to seem that the invisible creatures that so captured Leeuwenhoek might matter after all.

But it was chemistry, not microbes, on which Pasteur cut his scientific teeth. In 1847, almost twenty-five years old, he discovered that not all kinds of tartaric acid were the same. Two varieties had identical chemical compositions but a different molecular arrangement, making for an asymmetry akin to right- versus left-handedness. The chance discovery propelled him into a professorship at the University of Strasbourg.

By 1854, he was appointed dean of the Faculty of Sciences at the University of Lille in northern France, where he gained wider acclaim through a sugar-beet distiller, the father of one of his students. The distiller sought Pasteur's help in figuring out why fermentation sometimes failed, costing him thousands of francs a day. Pasteur visited the distillery and took samples from both the abnormal gray, slimy vats that no longer produced alcohol and from the normal foamy ones that did. Returning to his lab, Pasteur put a drop of liquid from one of the foamy vats under his microscope. He saw tiny yellow blobs of moving yeast. When he examined a drop of water from the other vat, he found no yeast swimming about. Instead, he saw gray specs filled with even tinier rods—organisms much smaller than the yeast, but just as obviously full of life.

The pattern was consistent. Samples from the abnormal vats were always acidic and contained tiny rods. He inoculated a sterilized flask of liquid with a drop of water from a vat filled with gray slime. Within hours, the rods multiplied and the vat turned acidic. Pasteur concluded that the rods (bacteria) made the acid, and that the yeast in the foamy vat somehow transformed beet sugar into alcohol. He decided that yeast and bacteria fought one another. A vat in which the yeast won produced alcohol—one in which the bacteria won turned acidic. Pasteur didn't describe it as such, but he was observing microbial ecology in action.

More important to his culture and day, Pasteur had solved the ancient mystery of fermentation: yeast did it. Microscopic organisms worked like little magicians, changing matter from one state to another. Captivated by his discovery, Pasteur became convinced that other microorganisms did equally amazing—and useful—things. He was certain that yeast fermented barley to produce beer, and grapes to make wine. When further experiments confirmed that yeast did indeed produce alcohol,

Pasteur was elated. All the wine in France and all the beer in Germany was not made by men, but by the labor of creatures too small to see. For this astounding revelation, the Academy of Sciences, which previously had refused to grant him membership, bestowed upon him the Prize for Experimental Physiology.

Unlike Leeuwenhoek, Pasteur was not shy about trumpeting his accomplishments. He boasted of discovering that microbes made meat go bad. Although he was not the first to claim a microbial basis for rotting meat, his elegantly simple experiments were convincing. He sealed one set of flasks partially filled with milk and sterilized them in boiling water, leaving another unheated set of sealed milk flasks at room temperature. After letting both sit for three years, he opened them.

In the unsterilized flasks, all the oxygen had been consumed and the milk had soured. In contrast, the milk in the sealed and boiled flasks was perfectly preserved. No microbes meant no decomposition, which meant no souring of the milk. The larger implications of his discovery dawned on him. Microbes were essential to life, central to nature's grand design. A world without microbes was a world in which nothing would decompose.

But where did they come from? The then-popular theory of spontaneous generation held that microbes sprang directly from dead and decaying organic matter. An alternative idea posited that life came only from life. A devout Catholic, Pasteur favored the latter. He believed life was created only once—back at the beginning of time. It could not just spring into existence any old day.

To test the theory of spontaneous generation, Pasteur filled glass flasks with a sterilized, and therefore dead, yeast solution. He then sealed the flasks so that no microorganisms could enter. If anything grew in the sterilized vessels, it would have to come from the dead yeast cells. Months later, he opened the flasks. They were free of life, a clear demonstration that discredited the theory of spontaneous generation. Microbes did not spring whole from decaying organic matter, they were spun from the fabric of life.

Eager to impress the French public with the practical advantages of supporting scientific research (like his), Pasteur entertained visitors with his uncanny ability to diagnose spoiled wine. Knowing that microbes were responsible for fermenting grape juice into wine, he blamed spoiled wine on microbial miscreants. He asked friends to bring him bottles of wine gone bad—bitter, oily, or starting to spoil into vinegar (ropy). Looking through his microscope, he found bitter wine infested with one kind of microbe, oily wine with another, and ropy wine—true to its name—filled with microbes all strung together.

He then challenged winemakers to bring him bottles of wine gone bad, confident he could properly diagnose problems without tasting a drop. Skeptical vintners accepted the challenge, conspiring to fool the naïve academic. They mixed bottles of perfectly good wine among bottles of sick wine to unveil Pasteur for the charlatan they were certain he was.

The growers laughed to themselves as Pasteur used a slender glass tube to pull a drop of wine from the first bottle and place it on a glass slide beneath his microscope. After a few minutes, the professor rose from his stool, turned to his smirking audience, and declared there was nothing wrong with the wine. When the taster, there to judge Pasteur's diagnoses, confirmed his opinion, the crowd quieted.

Doubt turned to amazement as Pasteur marched along the line of wine bottles, peering through his microscope at a single drop from each to correctly diagnose its condition: bitter, oily, ropy, or fine. Pasteur's performance impressed his audience. Invisible creatures could make or break the fruits of their own labor.

Not content to simply entertain, Pasteur worked to figure out how to keep microbes from spoiling healthy wine. If he could pull off this trick, it would show the French public the industrial value of his discoveries. He found that heating wine right after it finished fermenting would kill the microbes and prevent spoilage. Today, we call this remarkably simple and effective trick pasteurization.

Word of Pasteur's astonishing success at diagnosing the cause of spoiled wine interested the vinegar makers in Tours, who were plagued by vats of wine failing to turn into vinegar. When Pasteur inspected their kegs, he noticed a scum on the surface of the ones turning to vinegar. According to the vinegar makers, this always happened if vinegar formed and was absent if it did not. Inspecting a sample under his microscope, Pasteur found the scum was a thriving community of microbes happily oxygenating alcohol to vinegar. He then taught the vinegar makers how to cultivate and care for these tiny living chemists. Ancient secrets were being revealed one by one. Microbes could work for us or against us.

As it turns out, Pasteur saw only part of the picture when it came to how microbes influence wine. They also influence the different character of wines. Vinters generally attribute *terroir,* the "sense of place" that imparts distinctive flavors to wines, to regional soil, geology, and climate. While these differences certainly matter, the microbial communities in the early stages of fermentation vary in wines from different regions. This suggests that *terroir* reflects the influence of variations in the microbial communities associated with grapevines growing under different conditions. The next time you are savoring your favorite wine, you might consider thanking microbes for some of its distinctive character.

SHAKING THE TREE OF LIFE

More than a century after Pasteur's revelations, biologists were still unable to get their arms fully around microbes. Ever-more-powerful microscopes helped, but only to a point.

Then in the late 1970s, another Carl found himself taking a crack at organizing life. Carl Woese, however, declined to use the standard natural history approach of Linnaeus—closely observing and comparing external forms and features—still in favor at the time. Many a microbiologist of Woese's generation had parsed different shapes and features of bacteria to classify them. After all, some looked like spiny puffballs, while others resembled a piece of string, a chain of pearls, a kidney bean, or a curlicue. Some bacteria had flagella, tail-like appendages they used to propel themselves. Other bacteria lacked flagella.

Microbiologists used such features to sort bacteria into the taxonomic groups nested beneath the kingdom level—phylum, class, order, family, genus, and species. At the time, there were five kingdoms and all bacteria were in the Monera Kingdom. The other four kingdoms were the Protista (mostly microscropic life like amoebas), Fungi, Plantae, and Animalia. But looks are deceiving when it comes to microbes, and using the relatively few external features of bacteria was not a reliable way to classify them.

Woese was an unlikely person to take on the problems that microbes posed to conventional taxonomy, for he was neither a microbiologist nor a taxonomist. He began his career in biophysics not long after Watson and Crick received a Nobel Prize in 1962 for their work identifying the double helix structure of DNA.[1] At the time, biologists were still working out how DNA orchestrated the building of proteins inside cells. They knew that protein synthesis was linked to the sequence of base molecules that contributed to the structure of the much larger DNA molecule. But the "genetic code" had not yet been cracked.[2] Woese was convinced that the structure of DNA itself held important clues about one of the biggest, ever-simmering unknowns in all of biology—evolutionary relationships. To pursue his idea, Woese waded into the world of microbiology.

The genome of bacteria is relatively small compared to those of other organisms, which led Woese, and a new postdoctoral fellow in his lab, George Fox, to begin poking and prodding at its DNA. In particular, they focused on a gene called 16S ribosomal RNA in bacteria and abbreviated as 16S rRNA. This gene occurs in all organisms, and for good reason.

The 16S rRNA gene is needed to make ribosomes, which are small spherical structures found by the millions inside of all living cells. Ribosomes build proteins. And every protein found in every living thing—archaea, bacteria, protists, fungi, plants, and animals—comes spitting out of a ribosome the way fabric flies off a loom. No loom, no fabric. Proteins lend life much of its fundamental physical structure and so you'd *never ever* want the gene that makes ribosomes to change much. To do so could mess with the protein factories that make life.

But, there is a portion of the 16S rRNA gene that is not so stable, and it differs not only from bacterium to bacterium, but among all life forms. In the realm of bacteria, Woese and Fox believed that if they could nail down and document the degree of divergence in the variable part of the gene they would be able to better distinguish one bacterium from another. In addition, they thought that differences in the variable part of the gene could reveal how long ago the two bacteria had shared a common ancestor. In other words, using what Woese referred to as a cell's internal fossil record, they could construct something akin to a bacterial family tree.

But there were a few problems. They needed lots of different bacteria to sample; they needed to extract the 16S rRNA gene; and they needed to find a way to analyze and view the sequence of the base molecules in the variable part of the gene. Then there was the question of how to do these things.

Over days, weeks, and years, Woese and his lab tried out different techniques to catch a glimmer of the molecular structure of life. They landed on a way of analyzing the variable part of the 16S rRNA gene. Each DNA sample had a different molecular weight, which manifested as distinct, dark, blotchy bands once transferred to film and viewed over a light box. Woese himself hunkered down in his lab and scrutinized minute differences in the X-ray-like images of the blotches, over and over, thousands of times. He could link these differences to the sequence of base molecules in DNA. Painstakingly, this was how Woese and his lab worked out the sequence of base molecules for a small portion of the variable part of the 16S rRNA gene for each of their bacterial samples. They also applied the same method to DNA samples from multicellular organisms.

Seeing more than mere images, Woese began to read a new story of life. He turned the blurry bar-code-like blotches into a catalog of gene sequences. By 1976, Woese had become wickedly good at deciphering the blotches. He could recognize distinctive patterns associated with different groups of bacteria he'd been sampling. Multicellular life forms also had unique patterns. The expertise he developed was as extraordinary as Linnaeus's ability to identify mystery plants based on a few diagnostic features.

Woese's catalog of 16S rRNA sequences grew thicker. He believed his method of wrangling bacteria into taxonomic groups was bearing fruit. He also realized that, rod-shaped or curlicue, round or oval, the story his 16S rRNA sequences told did not square with the Linnaean method.

Naturally, Woese was interested in getting as many different bacterial samples as he could lay his hands on to keep growing his catalog. For some time a colleague had been puzzling over *Methanobacterium thermoautotrophicum*, a bizarre bacterium that could live at 150°F in sewage sludge with no oxygen. Perhaps a peek at its 16S rRNA sequence would reveal more about this hearty creature? Woese obliged.

The blotchy patterns on the viewing film from the heat-loving sewage dweller were distinctly different from all the other bacterial samples. More confusingly, the sewage dweller's pattern more closely resembled the DNA samples from multicellular life. What was this thing? Surely something was amiss. So another batch of DNA was extracted to generate another set of images. The second set looked just like the first. The discovery launched Woese and his lab into the rare euphoria of a scientific eureka moment. This strange methane producer was in its own world. Though it *seemed* like a bacterium, it did not fit the mold of all the other bacterial samples. The tree of life began to shake.

Woese sampled more life and squinted at countless more blotches. The results confirmed what he thought. The sequence of base molecules in the variable part of the 16S rRNA gene did indeed reveal evolutionary relationships—among all life, not just bacteria. The next year, in November 1977, the *New York Times* put Woese, in front of his scrawl-covered blackboard, on the front page above the fold.

The discovery of big news from the smallest of creatures laid the groundwork for rancor and controversy among biologists, for it had opened up a twentieth-century *Systema Naturae* of sorts. Woese saw a new way to categorize all life, but most biologists smelled heresy. Many did not entirely understand how Woese and Fox had arrived at their conclusions. Of course, the naysayers had not pored over the patterns of life on Woese's viewing films, nor had they figured out a way to translate the information into evolutionary relationships. They had, quite literally, not seen what Woese saw in staring at thousands of images. Bitter and ugly arguments ensued. Woese returned to his lab to decipher yet more blotchy films of gene sequences.

By 1990, other biologists who had been conducting similar research confirmed Woese's discoveries. Vindication of his results launched gale-force winds smashing into the tree of life. It appeared that not only were all bacteria no longer bacteria, but the classification of *all* life was about to change.

Classic field naturalists and many modern biologists couldn't stomach the reorganization of life when Woese proposed an entirely new taxonomic unit, called the "domain," above the level of kingdom. But there was something even more unsettling about this business of domains—who would be filling them. Multicellular life was shoved aside and unicellular life moved up in the world, way up. Bacteria would hold down one domain. Other one-celled organisms, among them the sewage dweller that started the whole mess, would occupy a second domain called Archaebacteria. (This domain has since been shortened to Archaea to emphasize that they are not, in fact, bacteria.)

Heated questions flew. How could two domains of life we can't even see constitute the basis for restructuring how we name and classify *all* life? Preposterous! Besides, the third domain, called Eukarya, shoved a ridiculous amount of life, including us, under a single taxonomic banner. We would have to hold our noses and share our place with slime molds and amoebas, algae, fungi, and liverworts, along with fish, whales, earthworms, trees, birds, and our beloved furry friends. Yet we have skin—not slime, feathers, fur, fins, or scales! Most disconcerting was that our species would now teeter on a flimsy twig near the end of a minor branch on the tree of life.

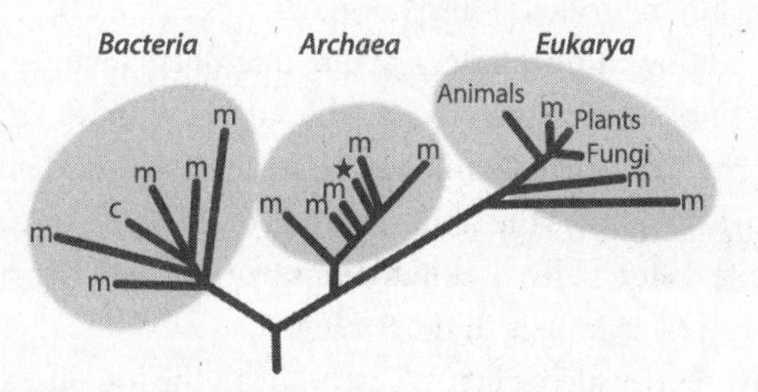

A New Tree of Life. The three domains of life and their evolutionary relationships. The star represents the group to which the sewage dweller belongs; the letter *c* represents cyanobacteria (early oxygen producers); and the letter *m* represents other groups of microbes (modified from Woese et al., 1990).

The griping and kvetching eventually died down and here is how we classify life today: the Archaea domain contains the most ancient organisms; the Bacteria domain contains regular old bacteria; and the Eukarya domain holds all the rest of life, including us. And this leads to a simple reality—two of the three domains (Archaea and Bacteria) and the new kingdoms within these domains contain one-celled microbes. And an entire kingdom in the Eukarya domain, the Protista, also consists of microbes. No matter how you cut it, microbes rule the roost on the tree of life.

Still, the idea of lumping all multicellular creatures together tugs at the part of our brain that clings to the idea that we are special. We and all our brethren in the Eukarya domain are collectively dubbed "eukaryotes" (you-CARRY-oats). Members of the one-celled Archaea and Bacteria domains are called "prokaryotes" (pro-CARRY-oats). These groupings are not taxonomic units. They are based on internal cell morphology, specifically whether genes float around freely within a cell (prokaryotes), or reside in a cell's headquarters, the nucleus (eukaryotes). We and all other eukaryotes have cells with a nucleus; prokaryotes do not.

Woese and many other microbiologists thought that the prokaryote-eukaryote division was a throwback to Linnaean ways. Why was the presence or absence of a cell nucleus any more important than another feature? Take an archaeon's cell wall. It is substantially different from a bacterial cell wall. And some protists (eukaryotes) have flagella, a common feature among bacteria (prokaryotes).

Further confounding matters about microbial identity is that the Woesean way of untangling who is who and how they are related also reveals that bacteria that should be the same based on their 16S rRNA sequences are not always the same. Bacteria of the same so-called species can lead quite different lives, akin to lions being able to live on lichens in the Arctic as well as impala in the Serengeti.

The chaos that ensued when Woese popped the top off Pandora's box to tinker with the tree of life doesn't clarify the situation with viruses, another group of microbes. Strictly speaking, viruses are not alive. They are neither single-celled nor multicellular, because they are not made of cells. Viruses are essentially small packets of DNA (or RNA) that come wrapped in a protein coat. And recall that viruses must reproduce inside of a living cell, where they usurp the host DNA to make copies of themselves. Some scientists think that certain viruses may actually end up helping or serving the host.

Now there is a movement afoot to add a fourth domain, a quasi-life domain of viruses. The rationale involves "giant viruses." These viruses are so big they dwarf the average bacterium. At one time the giant viruses were thought to be unusually large bacteria until genetic analyses proved otherwise. Sound familiar?

Whether or not the tree of life expands to incorporate a fourth domain, the fact is, most of life on Earth, however you measure it, is mostly hidden, invisible to our searching eyes and prying hands. At the heart of our confusion about how to wrap our big brains around microbes is that they still confound us. Are they friend or foe? With us, against us, or something else entirely? The more we learn about microbes, the more they seem to defy our attempts to categorize them. The latest discovery of a whole new proposed phylum of archaea called Lokiarchaeota (named after the mythological Norse trickster Loki) sheds light on the emergence of complex life.[3]

For most of the past century and a half we saw microbes as foes. Early in the development of microbiology, disease-causing organisms were emphasized for the obvious reason that if we wanted to vanquish them, we had to know who they were. Fortunately for humans, many of the major bacterial and viral pathogens that afflict us are readily culturable—meaning they can be grown and studied in the lab. And once a pathogen is culturable, it means, in most cases, that a cure or vaccine is close at hand. As valuable and necessary as it was (and is) to culture pathogens, it greatly biases how we think about and study microbes.

When gene sequencing became easier and faster during the 1980s, it gave scientists a way to study many bacteria and other microbes that had remained mysterious, since they couldn't be cultured. And once scientists started employing gene sequencing, they were in for a shock. The old methods of growing microbes in a lab had hidden much of the entire microbial world from us.[4]

Think of it this way—what would we know about the forests of the Amazon if we could only see and study one out of a hundred tree species? What would we know of marsupials if all we could see was *Vombatus ursinus*, the common wombat? If biologists after Darwin had realized the ecological context in which most microbes live, eat, reproduce, and die, they might have sooner recognized evidence for one of the greatest microbial feats of all time.

4

BETTER TOGETHER

One day, long, long ago, two microbes kicked off a remarkable sequence of events that forever changed the history of life. It all began when one of the most ancient of all organisms, an archaeon, merged with a bacterium. Their union created a composite life form, a microbial hybrid that jump-started the evolution of early unicellular life into more complex life. That's right, this chimeric life form eventually led to the likes of you, us, and every eukaryotic being that has, or will ever, walk, run, glide, squirm, slither, or swim across the face of the Earth.

Organisms living closely together, or within one another, are called symbiotic. The idea that microbial symbioses led to multicellular life had few supporters in the biological establishment at first. Most twentieth-century evolutionary biologists believed what Darwin had believed—evolution was a slow inexorable process of speciation driven by competition among individuals. But a persistent and brilliant scientist, Lynn Margulis, took on this traditional view of evolution in the 1970s and 1980s. She proposed a radically different evolutionary process based on ancient partnerships among some of the earliest types of microbes living on Earth.

Margulis, then Lynn Alexander, grew up on the south side of Chicago, the eldest of four girls, her parents neither academics nor scientists. But she was curious, driven, and maybe more than a bit mischievous. She got herself into a pickle when, unbeknown to her parents and school administrators, she left a high school–level program housed at the University of Chicago to attend her local high school. When school administrators discovered what she'd done, they wouldn't stand for it. She wasn't officially enrolled! Their solution was for Lynn to take a set of tests to see if she could qualify for early entrance at the collegiate level to the University of Chicago. Much to everyone's relief, she passed, and at age sixteen started down a path that would culminate in her becoming one of the most controversial biologists in America.

She graduated from college in three years, completely enthralled by biology and an older graduate student named Carl Sagan. Sagan helped her grasp the excitement and gratification of science. She and Sagan married, had two sons, and set off to the University of Wisconsin to pursue graduate studies. Carl continued in planetary sciences, and Lynn started a master's degree in genetics and zoology. The marriage didn't last, and Margulis moved on to a Ph.D. program at the University of California, Berkeley. There she continued studying the evolution of multicellular life, cultivating ideas that would rock strongly held beliefs in biology.

Back in the 1960s, when Margulis was a young graduate student, most geneticists saw DNA as paramount. It sat regally in the palace of the cell, the nucleus, issuing orders that governed most aspects of a cell's life. What went on in the rest of the cell was thought to have figured little in the evolutionary history of an organism.

Early in her graduate career Margulis discovered the largely ignored work of scientists from the late-nineteenth and early-twentieth centuries who had studied symbiotic relationships. Among them were the German biologist Andreas Schimper who, in 1893, proposed that chloroplasts (the photosynthetic organelles of plants) originated as bacteria; and the Russian botanist Konstantin Mereschkovsky who coined the term "symbiogenesis" in 1910. In addition, there was a fellow American, Ivan Wallin, who in 1927 described how bacteria might unite into new life forms, in his book *Symbionticism and the Origin of Species.*

When Margulis stumbled onto their ideas, they rang true. She found it plausible that certain structures inside of cells might have once been free-living bacteria. Such a newly forged partnership could have proved a better life strategy for either cell than living solo. At the time, few Western scientists considered the historical body of work on symbioses important, let alone that ancient partnerships between microbes might have led to life in the two kingdoms we know best—plants and animals. Where her peers saw little of interest, Margulis saw the biggest untold story in biology.

By the mid-1970s, the work of Carl Woese and others who were using genetic analyses to map out the tree of life began to support Margulis's view. Margulis, however, did not agree with Woese's new scheme when it was published in 1977, but not for the same reasons as most biologists. She thought the tree of life should reflect symbiotic relationships as well as molecular relationships.

While Woese had been looking at the 16S rRNA gene, Margulis had been thinking about and peering deep into cells, especially those of protists. She found herself embracing the out-of-favor ideas about symbioses first articulated a half-century before.

Pick up any introductory biology textbook and in the first fifty pages or so you're sure to find the classic cutaway three-dimensional illustration of a eukaryotic cell that reveals its internal organelles. The nucleus is the biggest part, insulated from everything else and tucked safely away in its own inner fortress surrounded by a special membrane. An interior ocean-like fluid called cytoplasm fills every cell. This miniature ocean ebbs and flows through canal-like partitioned areas formed by membranes that have folded in on themselves. Some organelles stay put and float in the cytoplasm; others go with the flow, moving around in the cell. The cell organelles look strange and have strange names—Golgi apparatus, mitochondria, and endoplasmic reticulum.

The work of a cell is not unlike a typical day for any organism. Certain things must happen for the cell to stay alive. It extracts energy from nutrients, excretes the wastes it produces, makes or repairs things, communicates with friends and foes, rests, and wakes up to do it all over again. To accomplish these things, molecules and chemicals are constantly shuttled around within a cell as well as into and out of it, bringing to mind a miniature bustling city.

Margulis thought that certain cell organelles—mitochondria and chloroplasts—had their own behavior and metabolism and seemed, paradoxically, to be part of, and yet not part of, the cell. What she pieced together largely confirmed the forgotten ideas of her predecessors. The way she saw things, eukaryotic cells and multicellular life arose from the physical union of free-living microbes.

It's surprising that Margulis chose to resurrect the idea of symbiogenesis at all. Despite the deep questioning of the status quo and the swirl of intense social and cultural change sweeping the country during the 1960s, Margulis found an entirely different attitude in the halls of UC Berkeley. She was shocked at how unaware her peers were of closely related fields. Paleontologists studied evolution, and geneticists were steeped in studies about the mechanics of evolution—chromosomes, genes, and DNA—but interactions between the two fields were essentially nonexistent.

Margulis referred to geneticists affiliated with what, at the time, was called the Bacterial-Virus Laboratory as "Johnny-come-lately biologists."[1] They had training in chemistry and physics, but not much in biology. They did not seem to know the first thing about the differences in cell division between nucleated cells (eukaryotes) and unnucleated cells (prokaryotes). At the time, Margulis was beginning to grasp the striking evolutionary implications of these differences for how parents pass traits to their progeny. The very manner in which she approached her work ignored the conventional scientific wisdom that specialization led to progress. And Margulis's view of evolution clashed with standard thinking on the subject in the Western world.

Was it a coincidence that a scientist from one of the most mercenary and capitalist societies of all, Victorian England, came up with the idea that competition drove evolution and that Russian scientists championed the idea of symbiogenesis? Whether one is a shopkeeper or a scientist, culture shapes the questions people ask and how they interpret what they see. Might not scientists from a communist country be more likely to study—or get support to study—life forms that appeared to be cooperating? With the Cold War raging between the United States and Russia, it was an impolitic time for Margulis to propose mutualistic partnerships as the key to the origin of higher life. At heart her view of symbioses—cooperation—as a key driving force in the history of life clashed with the Darwinian dogma that competition between individuals drove evolution.

Weak microscopic creatures influencing larger creatures? Bacteria banding together and driving the evolution of multicellular life, forming unholy truces with former enemies? Hogwash! Despite historical and cultural headwinds, Margulis gathered the threads of a story that would gel into an outrageous theory and endear her to few in the biological establishment.

In studying cells, bacterial and otherwise, and investigating their morphology and function, Margulis believed she saw evidence of an overlooked pathway for evolution. In 1967, after being rejected by fifteen journals, her radical idea—symbiotic relationships among microbes as the foundation for multicellular life—appeared in the *Journal of Theoretical Biology*. She was all of twenty-nine, with two young children at home.

Her bombshell idea was as stunning as it was controversial. She proposed that all multicellular life came about once single-celled life forms, predominately bacteria, physically merged. This strange and startling idea held that the evolution of higher life began when one cell ingested another and the preyed-upon cell did the unthinkable—it stayed alive. Margulis argued that symbiotic interactions and relationships were at least as influential in evolution as competitive interactions, if not more so. She called the theory "symbiogenesis," resurrecting the term from the earlier forgotten works that had so inspired her.

Margulis's championing of symbiogenesis did not go down well among her peers. Her critics considered symbiotic relationships as evolution's quirky side. Chief among them was Stephen Jay Gould, the renowned Harvard paleontologist. Gould's view of evolution was based on examining fossils and interpreting the rocks in which they lay. Focused on how environmental conditions and competition between organisms influenced the evolution and extinction of species, he didn't find much to get excited about in Margulis's ideas.

In his masterful 1,433-page book *The Structure of Evolutionary Theory*, published shortly before he died in 2002, little is said about microbes. There is one mention of "symbiosis," in the context of a short explanation of interactions that constitute biotic competition. Neither "symbiosis" nor "symbiogenesis" appear in the index.

The ways in which Margulis and Gould viewed evolution sprang from how they viewed biology in general. The different things they each saw—Margulis's emphasis on microscopic life and Gould's on plants and animals in the fossil record—led them to very different views of *how* evolution occurred.

Margulis believed that acquisition of whole genes and genomes through horizontal gene transfer (rather than small mutations within a single gene) had been critically important in the early evolution of life.

A one-celled organism like a bacterium that merged with another bacterium would double its genome. A multicellular organism, on the other hand, like a clam or snail, that acquired a new bacterium would have added but one new cell to its collective multitude.

This is entirely different than natural selection acting on variations in inherited traits. Consider the well-known example of the Galapagos Islands finches. Factors ranging from the size and availability of seeds to the presence of other finch species has been found to influence beak length and shape.

Gould studied broad patterns in the fossil record, not patterns inside of cells. Few paleontologists bought into Margulis's ideas. She continued to battle the headwinds of long-standing views about how evolution works. It was not only a case of biologists not seeing what she saw. They simply didn't see what she was saying.

Sexless horizontal gene transfer presented problems for conventional views of genetic inheritance, and thus the standard view of evolution. The idea of isolated pools of genes falls apart when you consider that the microbial world sips from a swirling stream of genetic material. Indeed, by some estimates only around 40 percent of the genes found in a named bacterial species typically occur in all members of that species. As for the remaining 60 percent of genes, they are variously present or missing in different individuals.

Stranger yet is that the closer one looks at bacterial DNA, the more dubious the concept of a species becomes. Bacterial genes, unlike ours and those of the plants and animals familiar to us, can change when circumstances like food sources or enemies change. Imagine a beaver acquiring a gene that allows him to fashion different sets of teeth when necessary—one for peeling the bark off of trees and shrubs and another to pick the flesh off the bones of a fish. Or imagine if your Labrador retriever could sprout fins or oversized webbed feet to swim after a tennis ball tossed into a lake.

But we're getting ahead of ourselves. Let's return to how Margulis saw symbiogenesis occurring over millions and millions of years. She believed that there was a specific sequence in which microbial life forms merged to create the forerunners of the nature we know—animals, fungi, and plants.

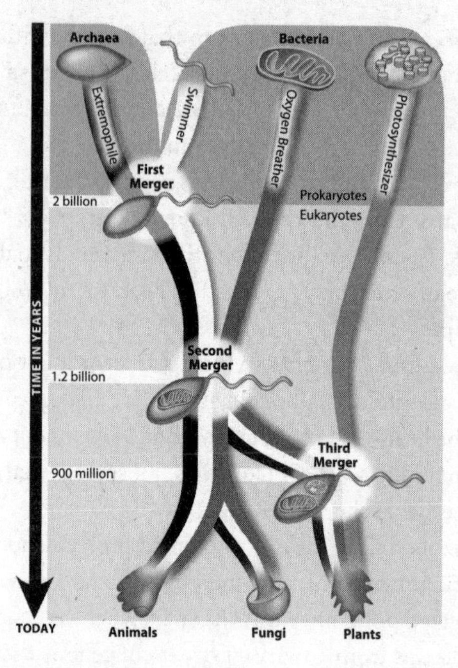

Microbial Mergers. Margulis's view of how an archaeon and a bacterium formed the first protist and laid the groundwork for all subsequent multicellular life. The second merger, in which a protist formed a symbiotic relationship with an oxygen-using bacterium, led to the forerunners of animals, fungi, and, later, plants (modified from Margulis, 1998, and Kozo-Polyansky, 2010).

The first merger involved a pair of ancient anaerobes, an archaeon and a swimming bacterium, uniting to create the first eukaryotic cell around 2 billion years ago. The earliest life, archaea had done quite well in Earth's early, oxygen-poor environments filled with mind-melting heat, bathed in ammonia and an abundance of hydrochloric acid and volcanically spewed sulfur. The archaeon's acquisition of a swimming bacterium (Margulis thought it was a squiggle-shaped bacterium called a spirochete) created a new organism. The swimmer became a tail-like appendage, propelling its new self around in Earth's vast seas. While the archaeon gained mobility, the swimmer acquired protection and a reliable food source (the archaeon's metabolic by-products). This first merger created protists, single-celled life forms, like the amoebas and algae that we call pond scum.

Almost a billion years passed, and then changing environmental conditions spurred a second merger. Atmospheric oxygen levels had risen because of the proliferation of photosynthetic bacteria and their waste product, oxygen, which allowed new types of oxygen-using bacteria to flourish. Naturally, the protist product of the first merger tangled with these new oxygen-using bacteria. When a protist ingested, but failed to digest, an oxygen-using bacterium, it created a new organism—a three-in-one job with the ability to live off of oxygen. This product of the second merger is the common ancestor of animals and fungi. Today we know the descendants of the captured oxygen user as the mitochondria that serve as the onboard cellular powerhouses of multicellular life.

Another microbial partner joined the roster of complex life around 900 million years ago when the now-established oxygen-using hybrid engulfed, but didn't kill, a cyanobacterium. Cyanobacteria had long harvested the energy of the sun through photosynthesis. Absorbing a cyanobacterium gave the archaeon + swimmer + oxygen-breather organism a solar-powered factory for making carbohydrates. This third microbial merger led to plants, and the chloroplasts that give plants their green color are the descendants of that first cyanobacterium. While details remain controversial, most biologists today accept the once-radical idea that microbial mergers led to multicellular life.

These ancient microbial unions set the course for the evolution of subsequent life. Lest you think these mergers sound upliftingly harmonious, Margulis described them as "violent, competitive, and truce-forming."[2] Despite the rough start, taking up residence inside of another cell provided the intruder with a safe haven sheltered from the dangers of the outside world. This kind of lifestyle offers clear advantages if you are a tiny creature surrounded by lots of bigger creatures eager to eat you. Teaming up also offered ways to take advantage of, or simply survive, the progression and accumulation of different physical environments on Earth. What Margulis worked out is that we, and every other multicellular life form, began a very long time ago as a series of symbiotic relationships between different microbes, mostly bacteria.

Viruses, however, are not a result of such mergers. Some biologists think viruses are damaged goods that arose when bacteria, in the face of intense radiation on early Earth, lost the defining qualities of life—

from containing themselves in a cell wall, to eating, excreting, and so on. Stripped down to the basics, viruses became little more than free-ranging packets of zombie DNA or RNA, with no option but to live and replicate within a host cell. While the question of how viruses arose remains quite controversial, they are far from inconsequential, as we would learn.

ASSEMBLING LIFE

Margulis considered the evolution of microbial life as a Tinkertoy-like process in which different forms of life built upon one another. She accepted that the accumulation of random mutations in genetic material occurred over time, but didn't agree that this was the be-all, end-all mechanism of evolution. The early evolution of multicellular life, as she saw it, was akin to collecting parts from different microbes to make a new organism. It would be like combining the drive train of a bicycle with the fan of a car radiator to make a pedal-powered fan—a new device with distinctive parts that retained the signature of their origin.

Margulis marshaled compelling evidence for symbiogenesis and its profound influence on the evolution of life. The cell organelles themselves held the key. She believed they displayed billboard-like clues to their symbiotic origin. To build her case, she turned to geology and the fossil record, the intricate imaging from new scanning electron microscopy, molecular analyses, gene sequencing, ecology, and biochemistry. However, if you rely mostly on the fossil record, or any other single line of inquiry to solve the mystery of where and what created the eukaryotic cell, you'd miss what screamed out at Margulis.

Remember all that eating, excreting, and so on in the life of a cell? Mitochondria power all this activity. And these biological powerhouses are one of the biggest clues to the symbiotic origin of multicellular life. They are thought to hail from the early oxygen-using types of bacteria that became incorporated into the early protist. The evidence? Well, for starters, mitochondria exist in all eukaryotes but not in any prokaryotes.[3]

In addition, the membrane surrounding mitochondria is unlike any that surrounds other cell organelles in its chemistry and function. This implies that, at one time, mitochondria led an independent existence outside of the cell. Mitochondria also have their own DNA and DNA replication process quite distinct from that in the cell nucleus. Moreover, they reproduce differently (via simple division) and not at the same time as the cell itself. To Margulis, this wide range of evidence all pointed to the same conclusion—long ago mitochondria moved into another cell and struck up a lasting partnership.

Chloroplasts, the descendants of the sunlight-utilizing bacterium that was the last to join the microbial merger train, exist today inside of every photosynthetic life form. Like mitochondria, chloroplasts have their own DNA separate from that found in the plant's cell nucleus. The DNA of chloroplasts is similar to that of cyanobacteria, the free-living photosynthetic bacteria we encountered earlier. However, the genome of chloroplasts is reduced because they have long lived protected inside of a host.

In the early 1960s, using electron microscopy that provided incredible magnification beyond that of compound microscopes, Hans Ris—one of Margulis's professors at the University of Wisconsin—was able to capture a detailed view of these so-called blue-green algae. He concluded that these cells were far less complex than a typical plant or animal cell. He thought they resembled the cells of *Streptomyces*, a soil bacterium. In other words, blue-green algae were and have been bacteria all along!

The third and still-controversial piece of evidence for symbiogenesis is the existence of structures, called cilia, in organisms that span all the way from modern-day protists on up to us. Cilia resemble tiny eyelash-like hairs attached to cells. Among the many functions of cilia are propelling some protists to their destinations and moving foreign material out of our lungs.

Margulis thought cilia arose from the merger of the swimming bacterium with a protist. She argued that cilia could not possibly have arisen randomly via gene mutation because of the structural similarity at the base of a cilium, which holds across innumerable life forms. Cilia must, she insisted, be one of the Tinkertoy-like parts donated by a bacterium. Although Margulis acknowledged that evidence for cilia having a previous life as a free-living microbe was weak, she believed it would just be a matter of time before additional evidence would be found that supported her view. However, the jury is still out on this point.

The implications of Margulis's work help explain the surprising, behind-the-scenes roles microbes play in the physical and biological worlds. In biology textbooks today, her theory of symbiogenesis appears along with diagrams of the cell organelles. And once-skeptical biologists are coming to see that symbiotic relationships abound.

The more scientists look into the relationships between microbial life and larger organisms, the more evidence they find for symbioses—in aphids, ants, coral reefs, and our own bodies. Recent discoveries have revealed that some microbes inhabiting specialized cells provide their animal hosts with amino acids the host cannot produce. Of course, there is the familiar union of algae and fungi to form lichens. But there are many more that are less familiar, like amoebas that practice bacterial husbandry, ants and beetles that protect and cultivate fungi, a sea slug that contains chloroplasts, and shipworms that host wood-digesting bacteria in their gills. And we are now learning that microbial symbioses are the norm for healthy plants and animals—part of why compost tea can help revive ailing plants and cows can live off indigestible grass.

What does the fossil record have to say about the role of symbioses in the subsequent evolution of life? Fungi came ashore long before plants did. And when plants joined them, this kick-started a collaboration that colonized the continents. Fungi probably evolved from protists that lived off dead algae. Dead and rotting plants fed fungi and fungi transformed unusable organic molecules, like the lignin in the cell walls of wood, back into nutritients that plants could use once again. The symbioses of plants and fungi define the two halves of a grand cycle that reuses nutrients—each half feeding the other in a symbiotic partnership that dates back to the earliest land plants more than 400 million years ago.

One reason symbiosis proved so successful as a microbial survival strategy has to do with the efficiencies it can offer. Imagine two different kinds of bacteria, each of which eats the waste produced by the other. They can sustain themselves as a community swapping nutrients back and forth in perpetuity. Neither takes over completely, but both persist indefinitely.[4] Over the long run, microbial communities with mutually beneficial checks and balances provide a more stable environment than individual microbes can find on their own.

It wasn't all that long ago that we figured out how we think nature works—at least the nature we can see, hear, and touch. The competition among individuals with distinct genetic traits—genes that win or lose along with the fate of the individual—explains a lot about the macroscopic world. But microbial ecology follows somewhat different rules in which competition often plays out at a population level rather than at an individual level. And this can favor symbiotic relationships among species in much the same way as villages and organized social life provided advantages to early human populations.

Working together, or in competition, communities of microbes alter the chemical and physical character of their environments. Each microbial species tends to do a limited number of things really well. Communities of collaborative microbes can do a much wider range of things than any one species can do on its own. Combinations of species that produce mutually favorable conditions thrive and persist. Those that don't, won't.

Ever since microbes teamed up to spawn multicellular life, cooperation and accommodation as much as all-out conflict shaped their relationships with plants and animals. Time and again, as the tree of life grew, relationships born in adversity became joined in necessity. Darwin would not have imagined that the microscopic world was such a collaborative place—or that some of the evidence is tucked away right inside of us. We inherited more than a third of our own genes from bacteria, archaea, and viruses.

The realization that microbial partnerships are both common and essential is reshaping how we see our relationship with the hidden half of nature. As these interdependencies come into focus, scientists are beginning to reassess earlier views that typecast microbes as the purveyors of disease—as threats to crops and people. In particular, we are learning how symbiotic relationships underlie plant health and soil fertility.

5

WAR ON THE SOIL

As our garden matured over its first five years, we stopped moving plants around and shifted our attention to building long-planned vegetable beds. The ideal place was the "back forty," the elevated area bordering the patio with a good southwestern exposure. A second round of weed-smothering cardboard covered with wood chips was starting to disintegrate into the dirt.[1] We needed to either move ahead with the food beds or raid another appliance-store dumpster for more giant pieces of cardboard.

Fortunately, we chanced upon a sale of broken pieces of rough-hewn granite curbing several feet long and six inches square. We hauled a few home. They were perfect for edging the new beds. We made a base of crushed rock and sand in shallow, narrow trenches and laid the granite in them to make a four-by-four-foot planting bed. Piling up the dirt from digging the trenches raised the beds to the top of the granite. An elevated bed meant that the soil would heat up sooner in the spring and hold more heat in the summer, helping plants grow better. Then we spread wood chips around the bed and had ourselves a garden within a garden. In a few months it looked like it had been there for years.

Back when our house was built, the back forty was where the builders piled up dirt excavated to make the basement. A jumble of rocks and topsy-turvy soil, it wasn't of the best quality, so we doused the bed every couple of weeks with soil soup and kept the ground between the plants covered with mulch. We set a dedicated bowl on the kitchen counter to collect scraps to feed the new worm bin. Several times a year we harvested rich, black vermicompost and mixed it in with the mulch around the plants.

Initially, the soil in the vegetable bed had been like everywhere else on our lot—putty-gray to khaki in color and on the clayey, rocky side. After a couple of years it turned the color of oily, dark-roasted coffee beans, still peppered with rocks but no longer lifeless. We saw creatures that weren't there before—beetles ambling through the mulch, earthworms, and sow bugs. At first we didn't give too much thought to how fast the soil was changing. We focused on the results instead—on eating what sprang from the soil.

Early summer in Seattle isn't like most places in the United States. We still get a lot of rain and sometimes it barely breaks out of the 60s. Seattleites complain about June-uary, but the cooler, wetter weather and light-filled (though not often sunny) days guarantee that lettuces and hearty greens thrive. All month we harvested a daily bounty of leafy greens. Tomatoes and squash arrived by August. Inspired by the fresh, delicious fruit of our efforts, we put in a second three-by-five-foot bed the next summer. More tomatoes, more lettuce, and Ozette potatoes grew in the new space. In no time at all the two raised beds produced a riot of edible plants. All summer we gathered ingredients for dinner right outside the back door.

In both the larger garden and the vegetable beds, the thing that came to impress us the most was how rapidly the mulch, worm compost, and soil soup altered the quality of the soil. Changing soil in such a short time wasn't supposed to be possible. Nothing we'd learned in college and graduate school suggested the potential to so effectively bypass the achingly long process nature goes through to make soil.

AFTER THE ICE

During the last glacial period, flowing ice stripped away whatever soil had been in Seattle. Once the climate began to warm, however, the glacier retreated north, leaving behind its legacy—the shovel-bending glacial till beneath our lot.

A glacier belies its inert appearance. Ice will flow like molasses if you pile it up high enough. The balance, or imbalance, between the accumulation of new ice in a glacier's source area and melt off at its snout controls how far a glacier reaches. More snow in the uplands and the glacier advances; more melting at its snout and the glacier retreats. All the while, a glacier works like a conveyor belt, transporting rocks and dirt until they drop off the front end of the melting ice.

Constantly on the move, glaciers crush and grind whatever they meet, compressing the sediments beneath into rock-hard glacial till. It's easy to imagine how glacial till forms if you turn a rocky road ice-cream cone upside down and let the ice cream plop onto the ground. The frozen cream and milk, like glacial ice, begin to melt and run off. Bits of chocolate, nuts, and marshmallow remain stranded in place, like the grains of sand, clay particles, and rocks that a melting glacier deposits. Next, put on the heaviest boots you can find and stomp the gooey mess into the ground. Then stomp some more, until you've smashed and mixed everything into the ground. When it all dries, you'll have an idea of what glacial till is like.

Once the glacier that overran Seattle retreated north, nature got busy making soil from scratch on top of the bare till and the sand laid down by meltwater streams. While Canada supplied the mineral component of our soil, the organic matter was homegrown. During the warm and wet postglacial climate, generation after generation of trees, shrubs, and ferns returned dead roots, trunks, branches, needles, and leaves to the soil. Dead animals—from microbes to mammoths—contributed as well. The result? Rich topsoil. Eventually, huge trees hundreds of feet tall anchored themselves and the soil snugly in place in what was to become our neighborhood.

All told, it took nature thousands of years to create the region's fertile soils, turning glacial till into some of the most productive soil in the world. The Pacific Northwest, at one time, was a vast temperate rainforest in which a single tree held more timber than it took to build our whole house. Then, a little over a century ago, in the late 1800s, a chorus of axes echoed through the ancient forests of Puget Sound. The engines of urbanization roared to life, carting off most of the fertile topsoil nature had so painstakingly made, stripping the soil down to glacial till again. It was as though a time machine swept through the city sending every newly cleared lot back to the end of the last Ice Age and the biology of a freshly poured sidewalk.

Think of the world's soils as impressions of Mother Earth's fingerprint, shaped by the climate, rocks, vegetation, and topography where they formed. No two regions have the identical set of ingredients for nature to work with. Prairie soils are carbon rich, tropical soils are oxidized and nutrient poor, and tundra soils are fertile but frozen. But no matter the location, there is one thing that every place on Earth has in common: soil is a two-way conduit through which the below-ground world that we can't see and don't know very well flows into our familiar, everyday world above ground.

Given how long it took our soil to recover from its glacial smackdown, and how long it generally takes nature to make soil anywhere, it was strange to think that we could so rapidly buck nature's trend in our little urban garden. We were not the first to marvel at the wondrous mystery of soil fertility. For most of humanity's agricultural existence, soil fertility was deified, personified as a goddess. Egyptians worshiped Sopdet, the Greeks idolized Demeter, and the Romans revered Ceres. Bad harvests were the price of angering fickle gods. For millennia people have recognized that their lives are linked to the soil. Even the biblical story of Eve and Adam poetically recognizes the dual unity of life and soil—the name Eve is derived from *hava*, life, and Adam is derived from *adamah*, soil. The origin of soil fertility remained one of the deepest mysteries long after humanity began farming and gardening.

GROWING IDEAS

In 1634, a Flemish chemist and physician, Jan Baptist van Helmont, began looking into the puzzling world of soil fertility and plant growth. This wasn't his first choice for how to spend his time, however. An alchemist by training, he believed that natural objects housed elemental forces that could attract and repel things—and could be understood through observation and experimentation. He ran afoul of the Church in rejecting a role for a divine hand in explaining natural phenomena. The unamused Inquisition condemned van Helmont to house arrest, charging him with impudent arrogance for investigating how God's creation—nature—worked.

Stuck at home for several years, he made the best of it and began thinking about how a tiny seed could turn into a large tree. How plants grew was far from obvious. Lacking a mouth and teeth, they don't chase prey or appear to consume anything at all. They just sit there getting bigger. Unconvinced by the prevailing idea that plants ate soil, he weighed and then planted a five-pound willow sapling in a pot with two hundred pounds of dried-out soil. Adding only water, he simply let the tree grow, the perfect experiment for someone imprisoned at home. At the end of five years, he reweighed the tree and found it had gained 164 pounds and that the soil had lost just two ounces. He concluded that the tree grew by taking on water.

Spurred on by his findings, van Helmont conducted a wide range of experiments. In one, he burned sixty-two pounds of oak charcoal, carefully collecting and weighing the resulting pound of ash and sixty-one pounds of gas (carbon dioxide). That burning wood produced ash was no surprise. But the production of gas, let alone so much of it, was a new discovery. Before this, the idea that most of a plant was fashioned from an invisible gas would have been laughable. If van Helmont had connected these two experiments, he might have realized that plants build themselves by combining water from the soil and gas from the air, along with a small amount of mineral-derived elements.

A century and a half passed before Nicolas-Théodore de Saussure, a Swiss chemist studying plant physiology, put it all together. In 1804, he repeated van Helmont's experiments, carefully weighing and accounting for the water and carbon dioxide that a plant consumed. A masterful experimentalist skilled with using new instruments accurate to hundredths of an ounce, he demonstrated that plants grow through combining liquid water with carbon dioxide gas in the presence of sunlight—a process we call photosynthesis.

De Saussure's discovery turned the understanding of fertility on its head. Plants did not draw carbon from the humus in soil; they pulled it out of the air! This reversal challenged the centuries-old notion that plants grew through absorbing humus (decaying organic matter). Still, de Saussure's work remained counterintuitive. After all, generations of farmers knew full well that manure helped their plants grow. As we'll see, this wasn't the only instance of a new theory overshadowing experience-based ideas about the origin of soil fertility.

While de Saussure's discoveries established that plants get the most basic elements for life from photosynthesis, this revelation left early botanists with another quandary—how did plants take up other essential elements, like the ones in van Helmont's ash?

LAW OF THE MINIMUM

The elements that make up plants start off in three places—the atmosphere, water, and rocks. Carbon (C) and nitrogen (N) come from the atmosphere and hydrogen (H) and oxygen (O) come from water. Rocks are the source for everything else.

Rocks form deep within the earth where incredible heat and pressure forge elements into a wide variety of mineral structures. Whether violently belched out of volcanoes or slowly exposed through the process of erosion, rocks begin to break apart once they reach the surface. Over time, extremes of temperature, water, and biological activity dissolve and pry rocks apart, releasing the elements they consist of into the environment.

Most rocks are composed chiefly of silicon (Si), aluminum (Al), and oxygen (O). The first two of these elements are so tightly bound to oxygen and locked up in mineral lattices that they weather out of rock into the soil in minute quantities. But plants don't need much silicon and aluminum. What they require a lot of are three elements—nitrogen (N), potassium (K), and phosphorus (P) to build roots, stems, and leaves. Earth's atmosphere is almost 80 percent nitrogen, and plants enlist microbial help to pull it from the air. Potassium is common in rocks. Phosphorus, however, is rare and only occurs in certain kinds of rock—and in decaying organic matter. Plants also need other major nutrients, like calcium (Ca) and magnesium (Mg), which are abundant in some types of rock and generally don't limit plant growth.

Other elements contained within rocks are considered micronutrients because plants need far less of them. Plants stockpile and concentrate critical micronutrients weathered out of rocks. Among these micronutrients are metals like zinc and iron, which plants fold into complex molecules that serve specialized purposes in their shoots, roots, leaves, seeds, and fruits. When we (or other animals) eat plants, the micronutrients in their tissues become part of our bodies.

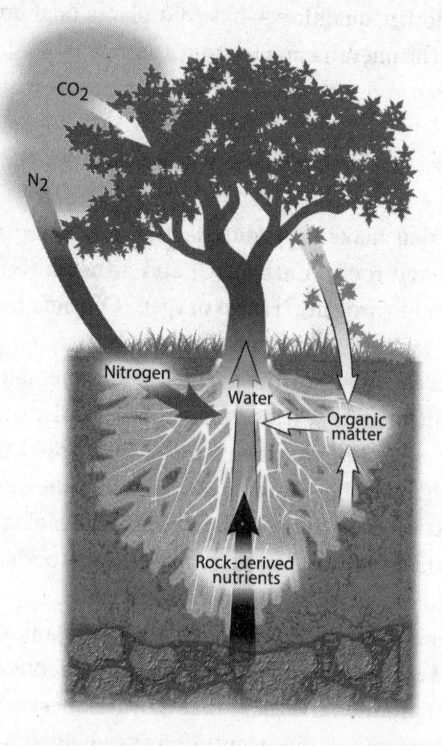

What Plants Eat. Plants obtain the major nutrients they need to grow from rocks, soil, organic matter, air, and water.

In something akin to a refining process, microbes help remove essential elements from rocks and keep them in play thereafter in the game of life. Microbes also recycle elements in organic matter. This has major consequences for soil fertility, as the remains of life hold a ready supply of nutrients in the right mix for supporting new life.

The study of plant health was in its infancy at the start of the nineteenth century. The route that elements took from rock to soil to plants wasn't yet understood. Neither was the function and importance of micronutrients to plant health. Not knowing the details, natural philosophers believed that soil organic matter, or humus—the thin dark layer at the top of soil beneath decomposing plant matter—somehow helped plants grow. The prevailing idea was that this mysterious material directly fed plants. Until, that is, experiments showed that humus would not dissolve in water, thereby discrediting the idea that plants could absorb nutrients directly from rotting organic matter. And if plants couldn't suck humus up through their roots, then how could they use it for growth?

Stumped, scientists of the day cooled on the notion that plants absorbed nutrients directly from humus. German chemist Justus von Liebig picked up the thread and led the charge on discrediting the humus theory of plant nutrition. In 1840, swept up in the Industrial Revolution, he wrote an influential treatise on agricultural chemistry in which he reasoned that the carbon in soil organic matter did not fuel plant growth because, as de Saussure had shown, plants obtained the carbon they needed from carbon dioxide in the atmosphere. Using then-standard practices of analyzing and weighing plant matter before and after burning it—Liebig found that plant ashes were rich in nitrogen and phosphorus. It seemed reasonable to assume that the matter left over in the ash was what nourished plants, and thus crops. This finding, in his view, provided the answer that plant scientists had long sought—soil chemistry held the key to soil fertility.

Liebig popularized his view as the "law of the minimum," a principle he borrowed without attribution from his German contemporary, botanist Carl Sprengel. The law of the minimum is the simple and still-accepted idea that the nutrient in shortest supply relative to the plant's needs limits plant growth. So, identify the limiting elements and you would know what to add to boost harvests.

In short order, Liebig and his disciples identified five key things essential for plants to grow—water (H_2O), carbon dioxide (CO_2), nitrogen (N), and the two rock-derived mineral elements, phosphorus (P) and potassium (K). They then jumped to the conclusion that organic matter played no important role in creating and maintaining soil fertility. By overthrowing the prevailing humus theory, Liebig ushered in a view of soil fertility at the heart of modern agriculture. Only late in life did he come to appreciate that decaying organic matter returned essential nutrients to the soil.

The appeal of Liebig's chemical philosophy is easy to understand when you read accounts of the explosive crop growth European farmers realized when they began fertilizing degraded soils with recently imported guano. In 1804, German explorer Alexander von Humboldt brought samples of this magic stuff back to Europe from an island off the coast of Peru, launching a nineteenth-century gold rush on fossilized bird droppings. In addition to containing a lot of phosphorus, this white rock held more than thirty times the nitrogen of most manure.

By the time the Peruvian guano islands had been mined into oblivion in the late nineteenth century, the widespread adoption of chemical fertilizers had become firmly entrenched as the guiding philosophy of agricultural production. When early-twentieth-century farmers found themselves working land that produced a fraction of the yields of their grandfathers', they flocked to new methods with the hope of reaping such harvests once again. It was a seductive thought—just add enough N, P, or K and crops would grow like crazy again.

Liebig's lasting influence ensured that agricultural science blossomed into a specialized branch of applied chemistry. The chemicals-as-soil-fertility mentality was the foundation for the work of specialists at agricultural experiment stations. Their research increasingly zeroed in on the individual parts of the soil, losing sight of it as a complex biological system. Soil biology and soil fertility came to be viewed as consequences of soil properties, rather than key influences on them. Agronomists saw soils as the product of chemistry, physics, and geology. Few considered soil biology important for increasing crop yields. And along the way crop yields became synonymous with crop health. Most researchers considered soil-dwelling organisms as pests to be managed or eliminated.

MICROBIAL WIZARDRY

The effect of nitrogen on plants is easy to see and near breathtaking for gardeners and farmers. It supercharges the growth of the green, leafy parts of a plant like nothing else. Once the guano was gone, finding more nitrogen presented a problem agrochemists were eager to solve. At the time, though, little was known about where organic nitrogen came from in the first place.

In 1888, two German chemists, Hermann Hellriegel and Hermann

Wilfarth, discovered microbes living in nodules on the roots of peas. They also noticed that peas and other legumes, unlike grains such as wheat, did not deplete nitrogen levels in the soil. Investigating further, they discovered that the microbes residing in the root nodules of beans, peas, and clover somehow converted atmospheric nitrogen (N_2) into ammonium (NH_4^+), which plants could take up and use. Hellriegel and Wilfarth had hit upon the secret to why the ancient practice of rotating crops of grains and legumes didn't rob the soil of nitrogen. A symbiotic relationship between certain bacteria and plants replenished it. Today we call this process nitrogen fixation.

You'd think that nitrogen would be easy to come by on a planet literally bathed in it, but plants cannot directly use atmospheric nitrogen. The triple bond cementing the two-atom molecular structure of atmospheric nitrogen is like a nearly indestructible lock, making it one of nature's most inert compounds. Plants can only use nitrogen after it gets cracked in half and combined with hydrogen or oxygen, creating one of two water soluble forms roots can absorb—ammonium (NH_4^+) or nitrate (NO_3^-). Because plants need a lot of nitrogen, and there's not much around in the right forms, nitrogen availability often limits crop growth in places where lack of water is not the more pressing constraint.

But there are only a few ways to get new nitrogen into soils and living things. The main way nitrogen enters the biological realm is through the bacteria that live in the root nodules of certain plants. Additionally, nitrogen-fixing microbes that do not form nodules have also been found in other plant tissues and in the soil. Recently, it has been discovered that some trees can get nitrogen from fungi that scavenge it right from nitrogen-rich bedrock. In one way or another, microbes drive almost all natural nitrogen fixation, except for a small amount made during lightning strikes powerful enough to crack the triple bond of nitrogen gas.

Once nitrogen is introduced into living things—as part of cells, tissues, and organs—it can cycle back and forth between the living and the dead. Soil organisms kick off the process of breaking down once-living matter, making it available to plants, and through them to us and other animals. Microbes convert the nitrogen derived from decaying organic matter in the soil back into water-soluble ammonium or nitrate. This process, called mineralization, makes the nitrogen available for plants to drink up in soil water. However plants get their nitrogen—from symbiotic bacteria or from recycled soil organic matter—microbes shepherd the process along. Over millions of years, this built up the stock of nitrogen circulating through the biosphere.

Despite the Hermanns' discovery of the nitrogen fixation process, the dramatic response of degraded soil to the application of guano and phosphates had firmly enshrined chemistry, not biology, as the foundation of modern agriculture. The microbial basis of nitrogen fixation soon seemed destined to become a quaint footnote to the march of progress.

As Europe industrialized at the close of the nineteenth century, concern over declining supplies of guano and rock phosphate led the president of the British Association for the Advancement of Science, the newly knighted Sir William Crookes, to focus his 1898 annual address on how to maintain agricultural production and feed the world. He urged scientists—and by this he meant chemists—to figure out how to bypass legumes and their microbial partners. Humanity needed to tap atmospheric nitrogen on an industrial scale. The only question was how. Crookes found an unlikely ally at the dawn of the First World War.

THE LAW OF RETURN

Nitrates were not just useful as fertilizer. They were essential components of the high explosives used in modern warfare. Lacking natural nitrate sources and therefore vulnerable to British naval blockades, Germany aggressively pursued new methods of nitrate production. In 1909, after years of failed attempts, laboratory chemist Fritz Haber successfully sustained production of ammonia (NH_3), the precursor for nitrate production. Another industrial chemist, Carl Bosch, rapidly commercialized Haber's process, and by the start of the First World War, Germany's new nitrate factory was pumping out twenty metric tons a day. At the end of the war, in 1918, all of Germany's synthetic nitrogen was going to munitions production, while the civilian population starved.

The discovery that industrial quantities of nitrogen could be produced opened a new era of intensive fertilizer use. It also earned both Haber and

Bosch Nobel prizes. That it required substantial energy inputs was of little concern at the time. Fossil fuels were cheap and plentiful. The way nitrogen fertilizer boosted plant growth on degraded soils seemed a scientific miracle. Synthetic nitrogen produced through what became known as the Haber-Bosch process is commonly credited with doubling crop production in the twentieth century. By the 1950s, the Haber-Bosch process surpassed biological nitrogen fixation. Today, about half the nitrogen in our bodies comes from the Haber-Bosch process.

The dramatic increase in crop yields cemented chemical fertilizers as the foundation of modern agriculture. Those who held alternative views of agricultural techniques were ridiculed or ignored by colleagues at experiment stations defending their specialized turf. It was obvious to all involved that the road to progress lay through specialization and industrial chemistry.

Almost everyone, anyway. On the far side of the world, an English agriculturalist took a very different approach to the problems of increasing crop yields and preventing plant diseases. Around the time that Haber and Bosch cracked the nitrogen problem, Sir Albert Howard was in the midst of discovering the rejuvenating effects of organic matter on soil fertility. But few wanted to hear what he had to say. Liebig and others had so thoroughly discredited the idea that humus fed plants that nobody in the agricultural establishment believed the Englishman.

Howard's eclectic career began in 1899 as a mycologist at the Imperial Department of Agriculture for the West Indies, where he specialized in diseases of sugarcane and cacao (the key ingredients in chocolate). Frustrated by a lack of land on which to experiment with his theories, he soon returned to England and took a position as a botanist at Wye College in Kent studying the diseases and insect predators of hops. Again, he encountered the same problem, no place to test out his many theories about why some plants thrive and others succumb to pests or disease.

In 1905, Howard jumped at an offer to become the Imperial Economic Botanist to the colonial government of India. The position held a key attraction, seventy-five acres of land on which to experiment as he pleased. He would be based at the Pusa Agricultural Research Institute, near modern New Delhi. There he became particularly interested in studying how changes in cultivation practices could boost crop yields and the response of plants to insects, fungi, and disease.

When Howard began his experiments on the fields at Pusa, he noticed that the crops of local subsistence farmers remained remarkably healthy and productive without insecticides or fungicides. Intrigued, he studied their practices and began copying what they did in his experimental fields. It was eye-opening.

> *By 1910 I had learnt how to grow healthy crops, practically free from disease, without the slightest help from mycologists, entomologists, bacteriologists, agricultural chemists, statisticians, clearing-houses of information, artificial manures, spraying machines, insecticides, fungicides, germicides, and all the other expensive paraphernalia of the modern Experiment Station.*[2]

For the next two decades, Howard continued his experiments, consistently arriving at opinions that conflicted with the beliefs of Liebig's disciples. Howard came to a radically new conclusion of why plants become diseased. He thought using pesticides and herbicides to protect crops from pests made it harder to grow healthy crops—and increased the need for more poisons. Insects and fungi were not so much problems as biological cleanup crews. They removed distressed or weak crops. In his view, modern agriculture was rushing headlong down a path that led to disease-prone crops.

This wasn't his only radical idea. Working at Pusa, he became increasingly convinced that the standard organization of agricultural research was counterproductive. Reports from agricultural research institutes were filled with the experiences of people working in separate fragments of science in segregated disciplinary fiefdoms (plant breeding, mycology, entomology, and so on). They were all, as Howard put it, "*intent on learning more and more about less and less.*"[3]

For Howard the proof lay in the fields. Despite increasing agrochemical use, a growing epidemic of crop diseases was shrinking harvests on British plantations. Disease-causing pathogens were increasing and spreading, not decreasing under chemical agriculture.

Howard believed that agrochemicals treated symptoms, not causes.

He thought that farmers needed a different strategy—understanding and supporting a plant's natural defense system. He got another opportunity to put his ideas into practice when the Indian Central Cotton Committee provided a grant to establish a new institute in Indore, an agricultural community in central India. Howard signed on to study the role of soil organic matter in relation to the problems underlying crop production. He was certain that microbes broke down organic matter and delivered nutrients critical for maintaining plant health.

Between 1924 and 1931, he developed a large-scale composting method called the Indore process, modifying a traditional agricultural practice of the area. The heart of the Indore process was in using both plant and animal waste to make compost. Howard set up large-scale field trials to test the method on cotton. The results were impressive—within a few years, crop yields more than doubled and crop diseases virtually disappeared from the fields. Plantation owners who gave it a try were impressed. Word spread and owners of large cotton, tea, and sugar plantations began to return organic wastes to their fields.

His colleagues at the agricultural experiment stations were another story. If composted humus held the key to improving soil fertility, plant health, and crop yields, as well as preventing the ravages of agricultural pests, then what use would there be for research on fertilizers, crop breeding, and pest control—the lifeblood of agricultural experimentalists?

It's not that the experts manning these stations sought to block progress. Far from it: progress was their mission, but they navigated with a different compass—Liebig's agrochemical philosophy. And from that perspective, the humus theory was wrong and Howard was way off base. While compost production might make sense for individual farm or plantation owners and their land, it was not a rational business model for an emerging industry that needed farmers and gardeners as long-term customers. Howard, in turn, considered the purveyors of short-term chemical fixes to complex biological problems the backward ones.

Plant pathologists also fanned fears that parasites would survive composting and wreak havoc on crops. They claimed pests would thrive on farms that relied on compost. After all, compost came from decaying plants and animal feces, sure sources of problems and pestilence. All in all, Howard had few allies in a world chasing the tail of technological progress.

While his foes sowed doubt and fear, Howard continued to conduct field trials. He reported on an experiment in which three acres of fungus-decimated tomatoes were removed and composted. Later the compost was applied back to the same field, which produced an excellent crop, free of fungal wilt. Similar trials conducted for other crops with other diseases proved that pathogens didn't survive composting. In his view the pervasive fears about composting were unfounded, plain and simple.

Howard's fondness for anecdotal evidence of real-life results from full-sized farms did not help advance his views in scientific circles. Neither did his vocal disdain for statistical analyses and small-plot experiments—the bread and butter of agricultural research.

Nonetheless, the roots of the modern organic agriculture and gardening movements stem directly from Howard's work. By experimenting with compost to restore fertility to tropical soils, Howard came to see chemical fertilizers as agricultural steroids, a way to enhance short-term performance at the expense of long-term soil fertility and plant health. He thought that Liebig's focus on agricultural chemistry had blinded the chemist. The fatal flaw in the new agrochemical wisdom, as Howard saw it, was that by focusing on agricultural chemistry Liebig and his followers overlooked the important role of organic matter. It fed the microbial and fungal catalysts that recycled once-living matter back into the building blocks of new life.

In contrast to his colleagues' view of chemical fertilizers as the foundation for maintaining soil fertility, Howard came to see the agrochemical approach as ensuring eventual disaster.

> *The slow poisoning of the life of the soil by artificial manure is one of the greatest calamities which has befallen agriculture and mankind.*[4]

Howard surmised that chemical fertilizers weakened a plant's biological defenses. He was confident the key lay below ground and suspected that fungal mycorrhizae were involved. In his view, the maintenance of soil fertility was the real basis of plant health and resistance to disease. The modern explosion of agricultural woes—pests, parasites, and pathogens—resulted from the breakdown of a complex biological system. Howard believed he'd found the secret to perpetual soil fertility.

Use both vegetable and animal waste (crop stubble and manure) to promote the growth of beneficial soil microorganisms. Farmers could ditch fertilizers if they took proper care of the soil. Howard's conclusions came from a lifetime of watching traditional approaches solve problem after problem created by agrochemical use.

Though long dismissed, Howard's general idea stood the test of time. In a 1937 appraisal of the condition of U.S. agricultural land, soil fertility was considered completely or partially degraded on about 60 percent of the total area under cultivation. Howard saw that synthetic fertilizers were becoming essential not so much to boost production as to make up for flagging soil fertility. In their zeal for fertilizers, farmers overlooked the potential to restore native soil fertility. And yet fertility could be maintained and even boosted by returning organic matter to farmers' fields. Harder than restoring degraded land, he realized, was changing the minds of the agricultural establishment.

Farmers often told Howard about declining yields in crop varieties fertilized continuously with chemical fertilizers. They complained that succeeding generations of crops gradually lost vitality and eventually the ability to reproduce. Agricultural authorities pooh-poohed such reports, pointing to decades-long field trials at the influential Rothamsted Experimental Station in Britain.

As early as 1843, amateur chemist John Bennet Lawes began experimenting with fertilizers at Rothamsted, his family estate north of London. Knowing that natural phosphates weathered out of rock far too slowly to be practical agriculturally, Lawes realized that treating rock phosphate with sulfuric acid would produce soluble phosphates that plants could take up immediately. He patented his process and established the first commercial fertilizer factory. The dramatic effect of his superphosphate fertilizer on crop yields delivered substantial profits that he used to turn his estate into a grand experiment on crop nutrition. There he established long-term research plots to investigate the effects of agricultural practices on crop yields. Lawes's main objective was to test whether wheat could be grown continuously with chemical fertilizers instead of farmyard manure.

By Howard's day, agricultural scientists viewed the long-running Rothamsted experiment as authoritative, the unassailable gold standard of agronomic research. On a tour of Rothamsted, Howard asked why official reports on the experiments did not mention the source of the seed for the crops. He was surprised to learn that each year wheat seeds were brought in from new outside sources. Every crop was grown from the best seeds available rather than from the prior generation of plants grown on the test plots.

Howard thought this made for a faulty experiment. Each October the plants got a fresh start. Few questioned the study's conclusion that fertilizers were effective at maintaining crop yields, but Howard believed they could not really evaluate crop performance over the long run if they brought in new seeds each year. Based on the experiences of real farmers, Howard was certain that the long-term effects of fertilizer use would have become apparent had seeds harvested from each year's crop been used for resowing the following season. In his view, the scientific basis for long-term, chemical-based farming was not as solidly established as its proponents trumpeted.

As it turned out, experiments carried out from 1843 to 1975 at Rothamsted showed that test plots fertilized with farmyard manure for more than a century *tripled* their soil nitrogen content. In contrast, nearly all the nitrogen added in plots treated with chemical fertilizers was lost from the soil, carried away in runoff or leached by groundwater. Eventually, the Rothamsted experiment showed that manure built soil fertility, whereas chemical fertilizers offered the quick fix of a temporary substitute. Unfortunately, the results came in too late to help Howard make his case.

But this didn't stop him from advocating for organic composting. Time and again, he saw, crop pests and diseases multiplied when farmers degraded their soil and then relied upon chemical inputs to prop up crop yields. And tellingly, he noticed that crop varieties that had been grown for centuries were being replaced with new varieties because yields of long-cultivated varieties gradually declined under intensive fertilization. On farm after farm, Howard watched his fears become reality as crops succumbed to disease after just a few generations on chemical fertilizers.

Inevitably, he saw farmers turn to chemical solutions to solve a biological problem, which then produced more biological problems. This vicious cycle resulted in greater demand for and opportunities to research, develop, and sell more chemical solutions, making pesticides the deadly handmaiden of fertilizers.

Howard saw that Western agriculture could not go on mining the soil forever, divorced from the wheel of life. Farmers needed to complete the cycle of life to refresh the land. He called this idea the Law of Return—the key to maintaining high crop yields was to return nutrients to the soil. Long-term farming had to be founded upon nature's principle of recycling life's hard-to-find ingredients. Soil fertility depended on the health of the soil microorganisms as much as the chemical makeup of the soil itself. Healthy, living soil was the key to soil fertility, plant resilience, and disease resistance.

Howard considered compost production essential to implement his Law of Return. The first stage involved using fungus to break down vegetable waste so bacteria could then process it into humus. Farmers could use whatever could be produced in the prevailing climate; that meant straw and hedge trimmings in Britain, sugarcane leaves and cotton stalks in the tropics. Animal waste was also critical, whether cattle or poultry—urine, dung, bones, or blood. Careful attention was required to make sure the compost was not overwatered, causing anaerobic conditions, or underwatered to the point where microbial activity was inhibited. Under the right conditions, several months of microbial action would turn organic waste into humus. The key was to culture microbes.

> *The essence of humus manufacture is first to provide the organisms with the correct raw material and then to ensure that they have suitable working conditions.*[5]

By the mid-1930s, methods refined in Howard's Indore experiment were bearing fruit. Plantation owners in Asia, Africa, and South America were reporting dramatic success in reviving worn-out lands. In 1940, near the start of the Second World War, Howard wrote *An Agricultural Testament*, a manifesto on organic agriculture. In it, he described what he learned studying the traditional agriculture of India and China, arguing that crop varieties fertilized with composted manure had been grown for centuries on the same land with no evident decline in productivity. He also noted that all manner of diseases are found among the plants and animals in nature, but they never assume large proportions.

He contrasted the remarkable lack of crop disease and damage from pests in Asia with conditions in the West, where the demand for pesticides and new, productive crop varieties skyrocketed after the introduction of chemical fertilizers. Unlike the effect of composted farmyard (and human) manure in Asia, which conferred disease resistance to crops, the reliance on chemical fertilizers in the West seemed to trigger a perpetual arms race against pests and crop diseases.

In *An Agricultural Testament*, Howard described the way the Inca created terraced slopes in the Peruvian mountains, sometimes with as many as fifty tiers rising up valley walls. Inca stonemasons built the outer retaining walls from large stones fit together as snuggly as the stones of the great pyramids of Egypt. After skilled craftsmen lined the terraces with clay, they regularly added organic waste to soil carried in from over the mountains. Their efforts turned barren hillsides into fertile terraced fields, some of which are still farmed today. Whether in Peru, China, Japan, or India, in each case the secret to generations of productive agriculture lay in the return of organic matter to the land.

Howard observed another differentiating factor in Asian agriculture. Typical farms of the Orient were small in size, which facilitated the return of compost to the fields. In 1907, barely a third of an acre of cultivated land supported each person in Japan. The 1931 census reported that the size of Indian farms averaged less than three acres. Leguminous plants also were commonly part of crop rotations in both countries. Long before Western science discovered the nitrogen-fixing role of microbes living in nodules on legume roots, experience had taught peasants around the world that peas, beans, and clover enriched the soil and organic matter helped sustain its fertility.

How could Western farming with its propensity for large tracts of land, monoculture plantings, and reliance on pesticides and fertilizers ever apply the lessons of small Asian farms? After all, even Howard acknowledged the allure of chemical fertilizers. They were easier to use on a large scale than farmyard manure. And tractors were superior to the horse in terms of power, and didn't require care, food, or rest, and fuel was dirt-cheap at the time. But tractors didn't produce manure. The impediments to Western farms adopting an Eastern agricultural practice were steep, but Howard thought the answer lay in industrial-scale composting. It all went back to his contention that farmers could ditch fertilizers if they fed their soil.

He didn't know exactly how compost worked, but he'd seen the effects of humus-rich soils over and over again in different settings. Even modest use of compost could result in rapid and marked improvements in plant growth. He realized that it wasn't simply the breakdown of organic matter itself. The fertility of the soil increased faster than the compost could decay on its own. Realizing that something else must explain a plant's dramatic response to compost, Howard suspected that it stimulated the relationship between mycorrhizal fungi and plant roots.

The idea seemed reasonable. After all, from his experience in India, he could see striking differences when he compared the root systems of tea plants treated with chemical fertilizers against those treated with humus compost. In fields infused with properly made compost, the roots were numerous and healthy. When examined under a microscope, they were teaming with mycorrhizal hyphae, delicate strands of the root-like parts of fungi. In contrast, the roots of plants in chemically fertilized fields were poorly developed and had few of the fine silk-like root hairs characteristic of healthy roots.

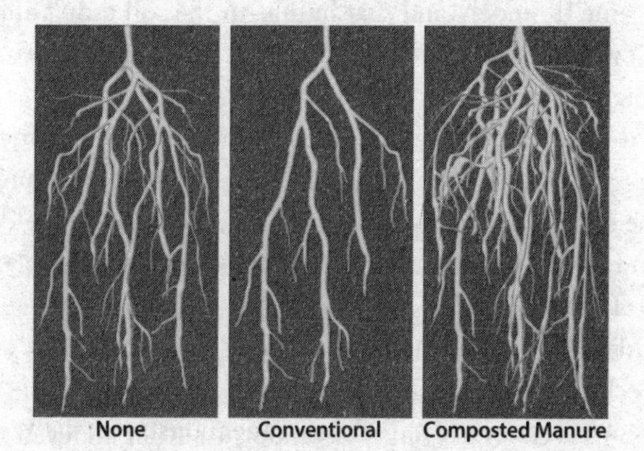

Root Health. Roots of hundred-day-old tomato plants receiving different fertilizer treatments (adapted from an image courtesy of Woods End Laboratories, Mt. Vernon, Maine).

Howard found further support for his view on a tour of tea estates in late 1937 and early 1938. In a test on tea seedlings at a plantation in Ceylon, two fields had been cleared of native humus and topsoil and stripped down to the subsoil. One field then received twenty tons of properly made compost per acre and the other received a standard dressing of NPK fertilizers. Nine months later, the tea plants in the compost-treated field were ten inches high, with strong taproots up to a foot long. They were well branched with abundant healthy foliage, their roots festooned with fungal hyphae. Plants in the field fertilized with chemical fertilizers were just six inches high, with shallow roots, and single, unbranched stalks sporting sparse, pale foliage. The compost-treated field better withstood drought than the fertilizer-treated field.

Howard thought that such conspicuous differences existed because mycorrhizae facilitated the relationship between plants and the nutrients held in soil. Fertility wasn't simply about the chemicals in the soil. It involved biological interactions among fungi, soil organisms, and plants.

> *Nature has provided an interesting piece of living machinery for joining up a fertile soil with the plant. . . . We appear to be dealing with a very remarkable example of symbiosis in which certain soil fungi directly connect the humus in the soil with the roots of the crop.*[6]

Howard came to understand that humus in the soil didn't affect plants directly; it worked through the action of microbial middlemen. This was what Liebig had missed.

Howard also believed he had found a mechanism by which chemical fertilizers increased the incidence of plant diseases among various crops. The destruction of the life in the soil, particularly the disruption of the fundamental relationship between mycorrhizae and plants, lay at the heart of the problem.

The widely held view among tea plantation owners that crop quality had declined since the introduction of artificial fertilizers mirrored Howard's own experience in the sugarcane industry in the West Indies. Before the advent of chemical fertilizers, mules and oxen played a role in maintaining crop productivity. The animals were bedded down with cane trash, and their manure mixed in and later returned to the fields. When cheap chemical fertilizers were introduced, the simultaneous arrival of tractors powered by inexpensive gasoline quickly made animals an expensive, outdated way to run a farm. The sugarcane crop itself, however, responded to a different bottom line. Insect pests and fungus took an increasing toll on harvests, sowing the seeds of increased demand for pesticides and disease-resistant crop varieties. Howard noted that certain large sugarcane plantations where only cattle manure was used had no problem with disease and little demand for new sugarcane varieties. India's cotton plantations had exhibited the same pattern.

Several years earlier, in the summer of 1934, Howard had returned to the United Kingdom. His house included a garden in which apple trees bearing poor-quality fruit were covered in the unsightly combination of American blight, green fly, and caterpillars. He decided to try building up the humus content of the soil. Within three years the pests were gone. The trees were transformed and produced first-class fruit. Again, Howard's results spoke for themselves. A healthy, living soil was the key to soil fertility, plant resilience, and disease resistance.

Even in the forests near where he had lived in India and England, Howard noticed how leaves, branches, and the dead trunks of fallen trees mixed with animal wastes and then decayed to humus through the action of microbial decomposers in the soil. Earthworms then reworked and distributed the humus. In this way, the forest supplied its own manure. Seeing the implications, he realized that even the most degraded soils could be revived—and maintained—with the application of well-made compost.

Howard considered worms both "*the gardener's unpaid handy man*" and canaries in the agricultural coal mine.[7] A thriving population of worms was a sign of healthy soil. Loss of worms foretold catastrophe. He described how the Connecticut Agricultural Experiment Station reported that the humus content of worm castings was 50 percent greater than that of the average topsoil. Worm castings held five times the nitrogen, seven times the soluble phosphate, and eleven times more potassium than the topsoil as a whole. Worms mixed soil with organic matter in their gut, and delivered it, freshly remade, back to the soil infused with plant nutrients. In essence, worms worked as miniature fertilizer factories, toiling day in and day out to manure farmers' fields.

Howard calculated that on good land worms produced at least twenty-five tons of nutrient-rich casts an acre. They did this for free year after year. How could it make sense to spray agricultural fields with chemicals that killed them? Howard saw earthworm farming as the way to grow thriving crops without the expense of chemical fertilizers. Feed the worms and you fed the soil.

Howard naïvely believed that the key to turning back the fertilizer industry's war on the soil was simply to show people what fertile soil could do for agriculture and gardening. He dreamed of converting Britain into a nation of compost gardeners. All across the English-speaking world there were derelict areas waiting to be restored. What better way to make his case than by returning to productivity the ruined soils of abandoned farms? The question was not what to do. Humus could restore farms surprisingly fast. The challenge was where to get enough organic matter.

Howard noted the tremendous potential for diverting back to the land the thirteen million tons of organic waste consigned each year to British dustbins. With such an effort, society could raise the fertility of soil in the fields surrounding cities. He proposed a vast salvage campaign to return vegetable, animal, and human wastes to the fields. The return of organic matter to the soil was no academic matter. At stake was the survival of agriculture—and thus civilization. Sewage should be filtered, composted, and turned into an "inoffensive powder" to be distributed to urban gardeners and farmers for use in restoring their soil. Neither his colleagues nor the general public were ready for visionary advice so far out of line with contemporary notions of progress.

During the Second World War, British Defence Regulations compelled farmers to apply chemicals to their crops. To help farmers, the government paid a portion of the costs, thereby subsidizing development of the fertilizer industry. The subsidies were not just about promoting better harvests though. A factory that made fertilizers could be easily converted to munitions manufacturing and vice versa.

The seeds for such conversions were laid at the end of the First World War. The Allies recognized the strategic value of the Haber-Bosch process and stipulated in the 1919 Treaty of Versailles that Germany share the secret of nitrogen fixation. The consequences rippled across the Atlantic, when in 1933, Congress authorized the Tennessee Valley Authority to dam the Tennessee River to generate cheap electricity. Among other things, the electricity would power fertilizer factories that could be converted to munitions factories. By the time Berlin fell near the end of the Second World War, electricity from these dams powered ten munitions plants. And, once the war ended, governments around the world sought new uses for instantly obsolete munitions plants. Fertilizer factories fit the bill—and preserved the option of converting production back to munitions on short notice. As swords turned into plowshares, the foundries of war found new markets in the war on the soil.

Howard's insights were trampled in the postwar drive for ever more mechanized and industrialized agrochemical farming. His work rapidly became overshadowed, collateral damage from aggressive government programs promoting fertilizer use to maintain munitions factories in waiting. The technical staff of fertilizer companies helped spread the new agrochemical gospel, assisted by the Ministry of Agriculture and a chorus of agricultural research stations and colleges encouraging farmers to pitch in and apply more chemicals to their fields. Crop yields became increasingly divorced from soil fertility—and the state of the land.

Yet we know today that Howard wasn't simply a grumpy agricultural Luddite. Overreliance on fertilizers, herbicides, and pesticides does foster vulnerability to agricultural pests. Due to their short generation time, agricultural pests spring back like weeds after a brushfire, as pests not killed off continue to breed and reproduce—their progeny increasingly resistant to pesticides and herbicides. And broad-spectrum biocides also can reduce or eliminate the competitors and predators that formerly kept pests and pathogens in check. In Howard's view, pesticides and fertilizers were as addictive as heroin. Quite effective at first, their efficacy plummets as more and more is needed to get the desired effect.

Still, many of those advising farmers and gardeners rarely question conventional views inherited from Liebig's agrochemical philosophy. You generally don't have to look far for a reason: they tend to work with or for agricultural suppliers that sell pesticides, herbicides, and fertilizers.

Shortly after we cleared our lot, Anne went to the Northwest Flower and Garden Show looking for ideas and inspiration. She visited the Scotts fertilizer booth and learned that, for no charge, they'd send technicians to our house and analyze our dirt. These commercial descendants of Liebig were eager to test our soil and sell us what it lacked. There in the booth, they laid out the program for our lawn—monthly doses of fertilizer and herbicides. As an incentive, we'd receive our first month free. Anne asked them about other ideas for what we could do to take care of our lawn. According to them, not much.

Unaware of how closely we were following in Howard's footsteps, we declined their offer, already having started down the road of returning organic matter to the soil. Our experience paralleled Howard's as we rediscovered for ourselves his insights into the catalytic effects of restoring life to the soil. All the organic matter Anne added to our soil built up microbial activity that helped support plant growth—which produced more organic matter. The surprisingly rapid positive feedback produced by organic matter and microbes led to a riot of plant growth.

In short order, bringing life back below ground triggered an explosion of life above ground. We saw what Howard saw—soil fertility springs from biology—all of the interactions between fungi, plants, and other soil organisms. Soil chemistry matters, of course, but how and whether plants can access nutrients matters just as much. The secret world of soil life, out of sight and long out of mind, is the key to building—and keeping—fertile soil. As we started looking into recent advances in soil ecology, we found they helped explain how and why Howard's ideas worked. Today the emerging view of soil ecology as the basis for soil fertility is not only undermining the chemical foundation of conventional agriculture, it is changing how we see nature.

6

UNDERGROUND ALLIES

Despite our very human draw to nature, we have a built-in aversion to rot. "Organic matter" sounds far more attractive than it actually is. After all, dead plants and animals look gross and smell bad. But once Anne and I saw with our own eyes what organic matter could do, those earthy smells seemed quite a bit sweeter.

Organic matter is the lifeblood of soil, the currency in the original underground economy. A soil's hunger for organic matter partially explains the mystery of why it disappears so quickly, but it doesn't explain how. Standing around above ground thinking you'll solve this mystery is futile. The action is underground. There, beneath your feet, microbes and larger life forms create complex and dynamic communities where everyone has a dual role—eat and be eaten.

Our darkening soil told us that all the organic matter we'd brought into our garden wasn't just disappearing into thin air. It was undergoing a transformation as if by alchemy. Each fall, Anne piled leaves on top of planting beds, and by the following spring they were well on their way to becoming one with the soil.

SOIL DOGS AND BUSY BACTERIA

The leaves (and wood chips and coffee grounds) in our garden mulch first attracted those with teeth or teeth-like equipment—the chompers, shredders, slicers, and tearers. These organisms do the work that a pack of dogs would do if given free rein in a room filled with bones, shoes, and fuzzy balls.

Most gardeners are familiar with some of the dogs of soil life—earwigs, beetles, and earthworms. They kick off the process of stripping useful nutrients out of organic matter until simple compounds of carbon, nitrogen, and hydrogen are about all that remain. Earthworms do impressive things despite their lack of teeth. Like birds, worms have a gizzard between their mouth and stomach that contains tiny rock fragments for grinding leaf mush and other organic matter into smaller and smaller particles. Earthworms suck in 10 to 30 percent of their body weight daily, mixing and blending organic matter and releasing the product into the earth. Just as in a cow's stomach, bacteria do a magical bit of work inside an earthworm, breaking down organic matter and rock minerals into nutrients for their hosts to absorb. What the worms don't use passes into the soil.

Smaller mites and springtails (visible with a ten-power hand lens) get in on the action. They are the ones that clip away all but the veins of a dead leaf, leaving their lace-like handiwork behind. All this activity shreds organic matter into microscopic pieces that, along with moist, nutrient-rich fecal pellets, make a hearty meal for fungi and bacteria. Unlike earthworms, microbes don't have teeth and jaws or a gizzard filled with tiny rocks. They don't need mouths. They use acids and enzymes to gnaw on organic matter.

These microscopic workhorses are busier than bees doing work in the soil similar to what happens in the rumen of a cow. They not only break down organic matter; they also play the role of supplier and distributor of nutrients, trace elements, and organic acids that plants need. So, while plants don't directly absorb organic matter, they do absorb the metabolic products of soil organisms that feed on and break down organic matter. Howard didn't quite grasp the details of this transformational feat, and for most of his life Justus von Liebig was satisfied with the notion that organic matter didn't matter. But now we know otherwise—soil organisms do the heavy lifting that keeps soil fertile and plants fed.

Fungi and bacteria anchor the menagerie of soil life and are the main brokers in this underground world. Fungi take the lead in turning dead things back into living things, and certain types specialize in eating dead wood, a particularly important role in nature's bustling economy. The chemical arsenal of fungi slowly breaks apart the long chains of incredibly durable carbon-and-hydrogen-rich molecules that form wood. Little else in the microbial world has the right stuff to dissolve the woody parts of plants. And whether digesting a log or a twig, fungi excrete metabolic products that feed other soil organisms that, in turn, become a meal for more soil organisms.

Although farmers have long despised fungi as vectors of plant disease, they are actually crucial to plant health and survival—and always have been. But like the microbes that cause infectious disease in humans, pathogenic fungi that harm plants steal the show. Only recently have we begun to appreciate the helpful varieties. Some symbioses between plants and fungi are so powerful that the one cannot live without the other. The famous truffles rooted out by trained pigs in France only grow with certain trees and thus cannot be farmed. Many more connections like these extend up from the ground and ripple through the biosphere.

The fecundity and tenacity of fungi are surprising given their delicate hyphae. These below-ground, root-like parts of mushrooms are so thin that it takes thousands packed together to reach the diameter of a piece of thread. More than once, Anne left a pile of wood chips sitting for a few months longer than planned, and when she finally plunged her pitchfork into the pile, she discovered golf-ball-sized clumps of what looked like cobwebs. Only the clumps weren't cobwebs, they were strands of fungal hyphae.

Scientists may consider fungi microbes, but hyphae are unlimited in how long they can grow. Forest networks of fungal hyphae are the largest life forms on Earth, forming subterranean networks that extend for miles. A single teaspoon of fertile soil can contain a half-mile of fungal hyphae.

Fungi secrete their wood-busting organic acids from the growing tips of hyphae, the way water drips from the nozzle of a hose. This acid bath helps further break down already tiny leaf and wood particles, allowing bacteria to get in on the action. And like fungi, bacteria further decompose organic matter with their own stew of chemicals. In the aboveground world it would take a set of grinding dog jaws or spinning blades of steel to do the work fungi and bacteria do in the soil. This dynamic duo is nature's janitorial and recycling service. Without them we'd all be neck-deep in dead plants and animals.

In the plant world even rocks serve as food so long as the right fungi and bacteria are in the soil. Turn over a stone on a forest floor and you may find the bottom covered with prospecting hyphae sucking mineral nutrients from bare rock. Some are particularly good at seeking out relatively scarce elements that plants need a lot of—like phosphorus.

A world of tiny middlemen lives between the microscopic realm of bacteria and fungi and the world of visible soil organisms. Like all middlemen, these organisms function as a pass-through, preying on each other and the primary decomposers—fungi and bacteria. Protists, like paisley-shaped paramecia, are one example of these middlemen. They propel themselves through the soil hunting for bacteria and nibbling on hyphae using rotating tufts of hair-like cilia. Amoebas too are protists. They direct the fluid held within their cell walls to make temporary structures called pseudopodia (false feet) as they blob their way through films of water on and between soil particles in search of food or on the run from predators. Nematodes, a diverse group of microscopic worm-like creatures smaller than a grain of sand, prey on protists as well as bacteria. These multilevel predator-prey interactions among microbes turn out to produce a key source of botanically available nitrogen—one of the big three plant nutrients. Bacterial cells are rich in nitrogen, which means that the poop of predatory protists and nematodes that dine on them is also nitrogen rich. This microbe manure can contribute a substantial portion of plant-available nitrogen to the soil—so long as bacteria are available as a food source.

When microorganisms decompose dead plants and animals, they put life's elemental building blocks back into circulation, including the big three—nitrogen, potassium, and phosphorus—and all the other major nutrients and assorted micronutrients important for plant health. Moreover, microbes deliver nutrients right back to where they are needed—a plant's roots.

Until recently, scientists thought that all organic matter in soil came from dead plant material. Walk around a temperate or tropical forest and this view makes sense. There are, literally, tons of plant parts lying around—seeds, fruit, bark, leaves, twigs, and branches. But researchers found that bacteria contributed a surprising amount of soil organic matter when they tagged some with a stable carbon isotope (^{13}C) and let them loose in an experimental plot for almost a year.[1] Subsequent soil sampling revealed scores of dead bacteria piled up on each particle of clay, sand, and silt. In fact, dead microbes accounted for up to 80 percent of the soil organic matter.

Now that you know this, you may never look at a pile of rotting leaves quite the same way again. Instead you might see a banquet table overflowing with food for soil-dwelling dinner guests who have the unique power to transform their meal back into nutrients that plants can use all over again. As every living thing in the soil eventually becomes something else's meal, an endless cycle of eating, dying, and pooping builds fertile soil from which new life springs.

Soil life digests and respires only about half of the carbon in organic matter, leaving the rest as dark-colored, decay-resistant carbon compounds known as humus. In our growing garden, it took only a couple of years before we began to see a thin layer of humus sandwiched between the mineral soil and the decaying organic matter on top. Fungi keep working away at humus to wring the last nutrients from the hardest-to-break-down bits of organic matter, like the fragments of durable waxes and resins from leaves and wood. Humus itself is so processed and transformed that it bears little resemblance to the original type of organic matter. While scientists find perplexing variability in its chemical structure, they agree that humus is a hallmark of rich, fertile soil.

The botanical world fed itself long before we came along. Plants drop their cast-off parts on top of the soil or, in the case of dead roots, within the soil. Hungry soil life eats this organic bounty and in the process changes death into the elements and compounds that plants need but can't get from photosynthesis. It is symbiosis on a grand scale. The hidden half of nature works the skin of the Earth, weaving a carpet of life that ripples up from the soil to the plants and animals that upon death become the foundation for a thriving microbial world. Life in the soil is the underground yin to the above-ground yang.

ANCIENT ROOTS

The way that our garden grew and changed mirrored the evolution of life on Earth. Nature bootstrapped herself up from the soil into the world above ground. Recent discoveries in paleontology and soil ecology reveal that soil ecosystems are far more ancient and every bit as diverse and complex as those that rise above them.

Microbes led the first wave of life that splashed ashore to colonize dry land at least 600 million years ago. But they left little direct evidence behind. Dating from the Ordovician Period (485 million to 444 million years ago), early evidence of soil life is blotches of color (or mottles), which indicate that bacteria modified organic matter. The first life forms adapted to tolerate desiccation and colonize land included bacterial mats, algae, fungi, and lichens. On their heels came stick-like, rootless, leafless wetland plants that could only reproduce in standing water.

Vascular plants evolved in the Silurian Period (444 million to 419 million years ago). The resulting supply of decaying organic matter fed the earliest land animals. Most were arthropods, invertebrates with an exoskeleton, segmented bodies, and jointed limbs. They were either detritivores that fed on rotting plants or grazers that fed on fungus or microorganisms. Fossils of arthropods, like predatory arachnids (relatives of modern spiders), centipedes, and millipedes tell us about ancient soil food webs. They consisted of photosynthesizers at the base, decomposers that recycled dead vegetation, and predators that ate microbes and detritivores. Arthropods have been remarkably successful for a remarkably long time. Today, their modern relatives—among them insects, arachnids, and crustaceans—make up more than 80 percent of all known living animal species.

Arthropods diversified along with plants and account for almost all land animals in the Devonian Period (419 million to 359 million years ago). Soil organisms mostly ate each other and dead, rotting vegetation. Early Devonian plant communities were quite simple, with a single layer of uniform vegetation less than six feet high. And as in a bizarre dream, giant mushrooms towering more than twenty feet tall grew on decomposing vegetation and bacterial crusts. Major interactions between plants and animals were indirect, through detritus feeding.

Detritivores contributed to soil development and cycled nutrients back to plant communities. Better developed, richer soils supported more extensive plant communities that provided shade and were more efficient at retaining water, which reduced the risk of desiccation. In return, thriving vegetation provided soil life with cover and food. This basic symbiotic relationship would prove lasting.

The advent of insect herbivory reshuffled animal food webs in the Carboniferous Period (359 million to 299 million years ago). Once animals began eating living plants instead of just dead or dying ones, two realms of life began diverging—the world above ground and the world below ground.

Through the subsequent evolution of plants—ferns, conifers (gymnosperms), and flowering plants (angiosperms)—the pattern set soon after life emerged from the sea stuck as the foundation for terrestrial ecology. Although the players at the top of the food web were shuffled several times by mass extinctions, the underlying symbiotic relationship between below-ground and above-ground life held. To this day, these two halves of nature remain tightly linked through the effects of soil life on both plants and the soil itself.

Throughout this evolutionary journey, plants continually came up with new structural and biochemical defenses to stay one step ahead of herbivores and pathogens. This ecological race shaped the subsequent evolution of insects and, later, mammals that dined on plants. It is one thing to appreciate the strategies that plants ginned up with each new foe that came at them above ground—caustic chemicals, thorns, and tough coatings—but consider the adaptive and selective forces at play between plants and soil life. Unlike fruits, flowers, and leaves, roots were a standard part of every plant early on. The race to stay ahead of below-ground enemies has been under way for quite a long time—since the beginning of terrestrial life, half a billion years ago.

We're only beginning to appreciate the specialized, ancient connection between a plant's roots and soil life. With time on its side, microbial soil life formed relationships with plants just as sophisticated as those between insect pollinators and flowering plants. Given the difficulties of observing what goes on in soil, we still have much to learn about below-ground relationships forged on the anvil of deep time. By some estimates, we still only know of about one out of ten soil-dwelling species. Until very recently the field of soil ecology was much like ancient astronomy, when our view was limited to the stars we could see with the naked eye.

A LIVING HALO

At the dawn of the twentieth century another German scientist made his mark in the world of plant research. Agronomist and plant pathologist Lorenz Hiltner made several key—and long-neglected—discoveries about plant defense. In 1902, he became the founding Director of the Bavarian Agricultural-Botanical Institute in Munich, the main purpose of which was to support agriculture in southeastern Germany. Like Howard, he doggedly pursued field trials to learn about the effect of microbial populations on plant health. Hiltner felt strongly that his research be useful and available to the average person and went so far as to write and distribute a pamphlet for gardeners or farmers, which at the time meant just about everybody in Bavaria.

In testing his unconventional idea that microbes could benefit plant health, Hiltner developed microbial amendments to promote plant growth, pioneering the study of microbial influences on plant nutrition. His experiments on potted plants showed that boosting the population of beneficial bacteria in the soil could counteract, and even reverse, declining plant health. Unfortunately, the First World War and the subsequent chaotic postwar period derailed his research.

Still, Hiltner's work established that the key to plant health lay, at least in part, with the microbes living at large in the soil. Like Howard, he recognized that despite the fact that all soils contain virulent plant pathogens, not all plants succumb to disease. His general notion, still accepted today, was that soils with high densities of nonpathogenic microbes confer growth and health advantages to plants compared to soils with low densities of soil life. This effect is so strong and universal that such life-filled soils are called "disease suppressive."

Scientists have now empirically and experimentally confirmed Hiltner's conclusions. In a classic experiment, replicated many times over, researchers grow plants in two types of soil. They sterilize one soil to kill all the microbes and they leave the other soil unsterilized. Then they introduce a known pathogen to each type of soil. Plants grown in the sterilized soil succumb to the pathogen while the plants grown in unsterilized soil do fine. That disease suppression results from microbial action is further demonstrated in another way. Sterilizing soil destroys its disease suppression, and on the flip side, mixing sterile soil with a tenth to as little as a thousandth of its volume with microbe-rich soil confers disease suppression.

As Hiltner intuited, the mechanisms responsible for disease-suppressive soils are linked to communities of microbes. Modern science is confirming that microbe-plant relationships are not one-sided affairs in which pests and pathogens call the shots. It turns out that when beneficial microbes are present in the soil near roots they send messages to plants that lead to an immune-like response called induced systemic resistance. What remained mysterious to Hiltner was the language nonpathogens use and how plants "hear" them.

Today we know that the words of the microbial language consist of a variety of proteins, hormones, and other compounds encoded by each microbe's genome. Plants use their roots as "ears" to listen to soil life. The net effect of this two-way communication is that it primes metabolic pathways that plants rely on to fend off pests and pathogens. Thus, in life-filled soil, plants stand ready to repel an onslaught, generally faring much better than plants unprepared to face an attack.

All plants have a microbiome, a cosmos-like mass of microbes that coats their roots, leaves, shoots, fruits, and seeds. Each plant hosts a distinct and unique microbial community. And if you are wondering if animals have microbiomes, they do. In fact, if you vaporized all life except microbes you'd be able to make out a shadow-like image made of microbes mirroring the inside and outside of all plants and animals.

Hiltner's major contributions to plant science focused on relationships between microbes, especially bacteria, and plant roots. He noticed that soil microorganisms were more abundant in proximity to plant roots, and gave this extra-busy zone a special name—the rhizosphere. Like a living halo, the rhizosphere surrounds every gossamer-thin root hair of a plant, millions of which project from every single root. The additional surface area that root hairs provide greatly expands the interactions of a plant with soil microorganisms. The rhizosphere is the richest part of a plant's microbiome in terms of diversity and numbers.

Although Hiltner didn't know it at the time, the number of microbes congregating in the rhizosphere is up to a hundred times higher than in the surrounding soil. He rightly hypothesized that the rhizosphere contained populations of microorganisms influenced by chemicals released from plant roots, and that the particular composition of the microbial community, in turn, influenced a plant's resistance to pathogens. Scientists have since discovered that unique, highly specialized communities of microorganisms do live in the rhizosphere. No doubt they live in the rhizosphere of every plant in our garden too. It turns out that plants recruit microbes to their roots using the oldest trick in the book—free food.

THE POWER OF FOOD

The amount of carbon in the soil greatly influences the abundance of microbial life. Plants pour carbon, in the form of carbohydrate-rich exudates, into the rhizosphere to feed the near-insatiable appetite of beneficial microbes. For the microbes it's as if someone has done all the work of getting food to the table, from growing and harvesting the crops to preparing and delivering meals. And it's easy enough for plants to do the work. After all, they can grab carbon straight from the atmosphere and make carbohydrates from scratch through photosynthesis. It's like printing money for the underground economy.

Exudates contain more than just carbohydrates. They are a stew of nutrients. Microbes in the rhizosphere also feast on the amino acids, vitamins, and phytochemicals found in exudates. You eat phytochemicals too. These are the molecules that lend distinctive colors and flavors to plants, such as the purple of an eggplant's skin or the sulfurous taste of Brussels sprouts and other plants in the cabbage family. Tobacco, one of the best-studied plants, produces over 2,500 types of phytochemicals.

When plants release exudates through their roots, bacteria and fungi flock to the rhizosphere. The offerings at the underground buffet go beyond exudates. Roots release mucus as they grow and slough off dead cells. To microbes in the rhizosphere these are yet more ready-to-eat carbohydrates.

Initially, scientists thought that exudates passively leaked out of roots, but then they looked more closely at cells on the periphery of roots. What they discovered were so-called border cells, filled with more mitochon-

Halo in the Soil. Plants secrete nutrient-rich exudates into the soil to draw beneficial microbes to their roots.

dria and internal membrane structures and vesicles than other cells. It turns out that these extra organelles allow border cells to help manufacture exudates and push them out of roots and into the rhizosphere. In other words, root cells do not lazily leak away resources. In their own way, plants are just as calculating and strategic as animals in getting what they need to survive.

When soil scientists discovered that plants release nutrient-rich exudates into the soil, they were astounded. One review found that root exudates can account for 30 to 40 percent of a plant's photosynthetic production of carbohydrates! That's like a farmer setting about a third of each harvest at the edge of the field for passersby to take for themselves. Why would plants give away such a bounty?

They don't. They trade exudates—for things they cannot make or do for themselves. The carbon fix that hungry microbes rely on doesn't come for free after all.

Seducing microbes with sugars and other substances might sound pointless, but it's the heart of the botanical world's defense strategy. Plants can't run or hide, but they have other defensive strategies, such as botanical swords (thorns) and shields (waxy leaf cuticles). Microbial recruits do the job below ground, taking on the role of palace guards to protect their botanical ally. Imagine a plant's root system as a castle in an underground landscape harboring microbial bandits and invaders. With a leafy hand, plants turn on a faucet and exudates start dripping down the castle walls and out into the rhizosphere. In this way plants use carbohydrates (and other compounds) that only they can make to attract and build a community of microbial bodyguards that displace, deter, or take out microbial enemies.

Plants are far from defenseless and vulnerable victims waiting around for pathogens to do them in. If, that is, the rhizosphere remains well populated with life that either benefits the plant or does no harm. In this case, pathogens have little chance of crossing the moat-like rhizosphere and breaching the botanical castle wall.

Factors such as temperature, moisture, and pH influence which nutrients become available to plants—and exudates help determine which microbes end up colonizing the rhizosphere of a given plant. Rhizosphere research reveals that substantial differences in the microbial species associated with plants growing in the same soil are attributable to the particular composition of root exudates that a plant pumps out. In other words, plants have the power to draw particular microbes to them.

There are several ways that nonpathogenic bacteria attracted to exudates help keep pathogenic soil fungi and bacteria at bay. The nonpathogenic bacteria, and to some extent fungi, generally consume exudates right away, thereby denying them to pathogens. And when beneficial microbes congregate on root surfaces, they enshroud roots in protective living cocoons, crowding pathogens out of the rhizosphere.

Some studies have identified another mechanism by which beneficial microbes keep pathogens away from the rhizosphere. Not just any old bacterium can sidle up to a plant's roots and hop on board. It turns out that differences in the genomes of symbiotic and pathogenic bacteria confer differences in how well (or not) each colonizes a root surface. Other studies have found that some species act as pathogens under some conditions and as symbiotic partners that cooperate with plants under other conditions. Genetic analyses may be able to discriminate between likely pathogens and symbionts, but microbial community dynamics are complex, influenced by interacting biotic and abiotic factors. Just as with above-ground biotic communities, the ecological role and importance of a particular species can depend on the company it keeps.

The microbes in the rhizosphere are an especially busy part of the plant microbiome. Different species of microbes have distinct preferences, and all are selective about what root exudates they'll take up. And they'll modify exudates too, for their own purpose, of course, but sometimes in ways that benefit the plant. For example, an amino acid exudate, tryptophan, takes on a whole new life when bacteria in the rhizosphere get ahold of it. Remarkably, bacteria can turn tryptophan into a plant growth hormone (indole-3-acetic acid) that causes roots to grow longer, sprout lateral roots, and increase the density of root hairs, all of which promotes the overall health of the plant. In this way, plants use microbial preferences to their own advantage, capitalizing on the fact that a different crowd will show up depending on what is available at the subterranean smorgasbord.

Given the multitude of interactions between plants and microbes in the rhizosphere, it's not surprising that scientists have discovered striking differences in the composition of rhizosphere communities in planted fields and bare earth. For example, pea and oat rhizospheres harbor up to five times more soil life than the rhizosphere of wheat or plant-free soils. And the pea rhizosphere is often particularly abundant with fungi.

The phytochemicals that plants release in root exudates are another botanical defense strategy. These compounds counter a wide range of above-ground and below-ground threats. Despite the distinctive flavor and health benefits that we reap from phytochemicals, plants make them for their benefit, not ours. Some phytochemicals are standard fare in root exudates, where they stimulate or inhibit the expression of bacterial genes, attract beneficial bacteria to roots, and deter root pathogens. Freshly germinated seedlings of some plants pump out sulfur-containing phytochemicals that foster the growth of mycorrhizae and bacteria. In some instances, phytochemicals act like traffic flaggers waving certain bacteria and fungi into the rhizosphere. Plants can also release phytochemicals that send a "wrong way" message to approaching microbes. If a troublesome microbe ignores the warning, it gets a stronger signal to back off. If ignored, the plant initiates a cascade of chemical defenses to seal off entry points.

Plants can also make antimicrobial compounds that kill or weaken pathogens. For example corn (*Zea mays*) exudes antimicrobial compounds (benzoxazinoids) in sufficient quantities to inhibit many soil microbes in the area immediately around the roots. Sometimes beneficial bacteria in the rhizosphere will produce metabolites that help to keep pathogenic fungi at bay. Of the three major groups of rhizosphere bacteria—Actinomycetes, Firmicutes, and Bacteroidetes—Actinomycetes, in particular, produce a wide variety of compounds that interfere with bacterial, fungal, and viral pathogens.

There are striking examples of plants enlisting help from bacteria living in the rhizosphere to fend off trouble from above. When leaf pathogens attack, the plant senses them and sends a long-distance chemical message down to root cells. The root cells, in turn, start releasing exudates. In one case, a very specific exudate, malic acid, acts like a shepherd calling his dogs. *Bacillus subtilis* comes running and within hours it densely colonizes the roots and initiates more chemical communication with the plant. The bacteria-plant conversation triggers the plant to produce and circulate systemic defensive compounds against the leaf pathogen. More amazing, *B. subtilis* also induces plants to shut down the tiny openings on leaf surfaces, called stomata, that pathogens use to slither their way inside a leaf.

Although the way that microbes help plants defend themselves is ingenious, they also play another, and equally impressive role in plant health. But only particular kinds of plants will take the first step in this carefully honed dance. Legumes, plants in the pea family, produce phytochemicals called flavonoids to attract nitrogen-fixing bacteria. Hiltner didn't know about flavonoids per se, but he realized that certain types of bacteria made nitrogen available to plants.

One of the most well-known groups of nitrogen-fixing bacteria is in the genus *Rhizobium*. These bacteria, also called nodule-forming, move into the rhizosphere when plants release flavonoids in their root exudates. Nodule-forming bacteria are very selective about what plants they will associate with. Only if the plant sends the right chemical message will bacteria respond with their own offering to the plant—special molecules called "Nod factors" (shorthand for nodule-forming factors). Nod factors work like an ID card, assuring the plant that the nodule-forming bacteria are who they say they are. So long as the plant keeps the welcome mat rolled out, the seduced bacteria lock onto a root hair and the process continues. The flow of Nod factors soon snake-charms the root hair cells to grow in curls and bulges around the bacteria. Safely tucked into their stable-like nodules, the bacteria become a nitrogen-fixing herd working for their host. In return, plants supply their new partners with a steady stream of food. If the plant needs more nitrogen, it secretes more of the appropriate flavonoid to recruit more bacteria. Over time, whole colonies of nodule-forming bacteria relocate from the open frontier of the soil into the safety of a root. The nodules that nitrogen-fixing bacteria inhabit are such an integral part of a plant that they are considered an accessory organ.

A phytochemical invitation and a Nod-factor reply is but one of many ways plants seek out beneficial bacteria and use them to either acquire nutrients or repel pathogens. In addition to the bacteria that live in nodules formed on roots, other species actually live *inside* of a plant and are called endophytes (*endo* for "inside" and *phyte* for "plant"). From crown to roots, such bacteria tuck themselves in between plant cells, where they release compounds that spur plant growth and enhance a plant's resistance to pests and pathogens. Such relationships appear to be ubiquitous in plants, but are well understood in only a few cases.

Legumes are not the only plants that rely on bacteria to meet their nitrogen needs. Alder (*Alnus* spp.), poplar (*Populus trichocarpa*), and willow trees (*Salix sitchensis*) recruit nodule-forming bacteria to help them colonize nitrogen-poor ground, such as bare gravel bars in a river. Over time these first plant colonizers drop enough leaves to build incipient soil, making it possible for other plants to become established and grow. Nitrogen-fixing bacteria have also been found living in the tissues and rhizospheres of important crops like coffee (*Coffea arabica*), corn (*Zea mays*), and sugarcane (*Saccharum* spp.).

The amount of plant-available nitrogen that nitrogen-fixing bacteria provide can be substantial, adding up to 200 pounds per acre each year, depending on soil conditions. This amount would be enough to offset chemical fertilizer for wheat and corn, which generally falls in the range of 100 to 200 pounds per acre. Growing in the stems of sugarcane, the nitrogen-fixing bacterium *Acetobacter diazotrophicus* can fix up to 150 pounds of nitrogen per acre, also enough to replace chemical fertilizers. Until the invention of industrially produced nitrogen fertilizer, almost all of the nitrogen in plants came from bacteria that converted atmospheric nitrogen into a plant-usable form. And once incorporated into a plant, bacterially acquired nitrogen cycled from plants to soil to animals over and over.

FETCHING FUNGI

Just as there are beneficial bacteria, there are also beneficial fungi. In fact, one of the chief factors contributing to disease suppression in soils at large is fungal diversity. And fungal symbioses with plants are far more ancient and pervasive than those between nitrogen-fixing bacteria and legumes. The ancient origin of such partnerships explains why mycorrhizal associations are known to occur in about 80 percent of flowering plants and all gymnosperms (nonflowering plants like conifers).

The chemical signaling between mycorrhizal fungi and plants is less understood than that of nitrogen-fixing bacteria, but involves similar interactions. Fungal hyphae have their own version of Nod factors called Myc factors. The communication between plants and mycorrhizal fungi is far more intricate than most scientists imagined just a few decades ago. Such chemical signaling has long been suspected, but the chemicals responsible have only recently been documented.

The first step in establishing the symbiotic connection between mycorrhizal fungi and plant roots is the onset of branching in the root-like hyphae of fungi to establish contact with the host root. Plants release flavonoids to initiate the branching. Mycorrhizal fungi are allowed to enter the courtyard-like space between cells of the botanical castle but no further. This level of access is enough to form a living bridge between fungi and plant. A delivery system doesn't get much better. It's like the postal service bringing packages inside your house and laying out the contents right where you want them. And a plant with ready access to the minerals and molecules needed to make phytochemicals and other defensive compounds has a leg up on pathogens.

At the opposite end of the hyphal network, symbiotic mycorrhizae extend their razor-thin hyphae out into the soil. Serving as extensions of root hairs, the mycorrhizal hyphae increase the effective surface area of the root system as much as tenfold, and provide access to small, otherwise inaccessible, soil pores and cracks. Mycorrhizae help plants wring more nutrients and water from the soil—doubling or tripling the uptake of phosphorus (and other nutrients) per unit root length. In return, the plants provide the fungi with catered meals of carbohydrates.

Mycorrhizal fungi not only help nourish plants directly; they help improve soil quality. Their networks of hyphae create soils far less prone to erode and greatly enhance water infiltration. These positive effects on soil improve plant productivity—and resilience in droughts.

The effect of mycorrhizal fungi on soil structure and nutrient acquisition are among the mechanisms Howard sought but couldn't find. It would certainly come as no surprise to him that such physical connections exist or that plants and mycorrhizal fungi exchange nutrients. This little-explored crossroads of geology and biology is the union of the subterranean and sunlit worlds. Here exchanges of nature's wares—the carbohydrates from plants and foraged mineral nutrients from fungi—form an underground economy.

What is clear to scientists today is that the flow of carbon and nitrogen in the rhizosphere is extremely complex and bidirectional, with roots both secreting and absorbing organic compounds. Roots exude ions, enzymes, mucus, and a wide array of organic compounds. Surprisingly, plant roots also take up organic acids, sugars, and amino acids. While the carbon uptake by roots plays a minor role in a plant's carbon budget, some plant roots turn carbon compounds taken up from the soil into organic acids that are then exuded back into the soil, where they improve plant uptake of phosphorus in the rhizosphere. In addition, some symbiotic microbes take up exudates and produce metabolites that, in turn, chelate iron and phosphorus from the rhizosphere, making these elements available to plants.

The bottom line is that the interactions between plants and soil life—especially bacteria and mycorrhizae fungi—are far more intricate than previously imagined. Plants actively push nutrients out into the rhizosphere to feed particular microbes that help protect plants from pathogens or usher key nutrients into their roots. In fact, the entire plant microbiome operates much like an ecological pharmacy for its host, helping to keep the currency of life circulating. Our deepening understanding of this connection between plants and soil life is akin to the evolution in thinking from Newtonian mechanics to quantum physics—the former a cartoon version of reality based on what we can perceive directly with our senses, and the latter a deeper story of the complex variability that underlies it all and explains how nature really works.

SILENT PARTNERS

Ongoing discoveries in soil science continue to unveil a new view of plant health that is changing the way we think about nature—and agriculture. For centuries, gardeners and farmers used compost, manures, and other organic nutrient sources to grow healthy plants, boost crop yields, and replenish soil fertility despite not fully understanding what was happening beneath their feet. And the great thinkers of the time, like Liebig, overlooked how soil life, working in concert with the chemical composition of a particular soil, contributed to soil fertility.

Adding fertilizers to soil doesn't guarantee that they'll make it into a plant. Without the help of microbes transforming nutrients into forms plants can use, important elements remain uselessly parked just outside a plant's roots, like a cargo ship stuck outside of port. The still-emerging view of the extraordinarily intimate relationships between soil microbes and plants paints a radically new picture of soil fertility. While plant nutrition and plant health certainly have a chemical basis, Howard's vision of soil fertility as a complex ecological system now seems prophetic.

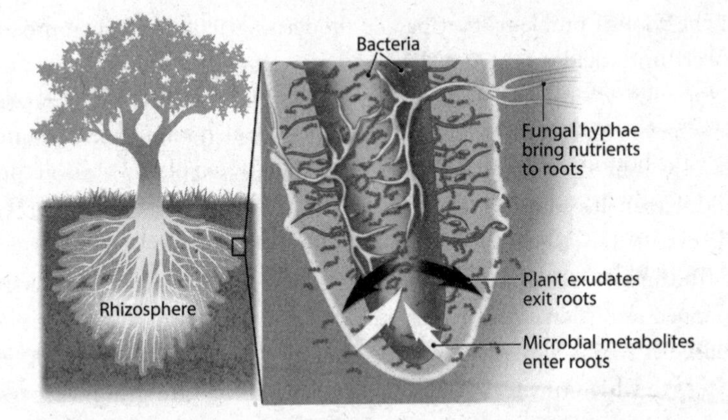

The Underground Economy. The rhizosphere around plant roots is the site of countless trades between plants and soil microbes. Both fungi and bacteria consume plant exudates, and in exchange they provide plants with nutrients and metabolites essential for growth and health.

Soil ecology is still in its infancy compared to the far more established fields of soil physics and soil chemistry. Only in recent decades have biologists begun to tease apart the major groups of soil microbes and their functions and relationships as members of subterranean communities. They've already learned that the marvelous coevolution we see between plants and pollinators is mirrored below ground in relationships every bit as complex but far less understood. Out of sight, and out of mind, beneficial microbes do heavy lifting that helps sustain plant health.

The new understanding of microbes as biological catalysts of natural soil fertility challenges the philosophical foundation of modern agriculture. No one can deny that agrochemistry has been effective at pumping up crop yields in the short run. But it increasingly seems that in so doing we endanger crop yields over the long run. In addition to disrupting nutrient transfer, overuse of agrochemicals can degrade and disable plant defenses, opening opportunities for pathogens to attack vulnerable crops. By inadvertently decimating beneficial soil life, we disrupt the nutritional and defense systems plants have honed through adaptive symbioses with microbial life.

If you think about soil as a biological system, it's easier to grasp why an agrochemical approach that "manages" a few plant pathogens lies at the root of problems that plague modern agriculture. When broad-spectrum biocides take out the good as well as the bad, it is the bad actors and weedy species that bounce back fastest. This fundamental flaw makes biocide-based agriculture addictive—the more you use the more you need. While this makes good business sense for dealers and middlemen, it's counterproductive for the user over the long run. And in the case of agriculture, that means all of us.

The importance of coevolutionary relationships below ground recently snapped into sharp focus for us halfway around the world. On a trip to southern Africa while writing this book, we were surprised to see the extent to which tiny termites influence the whole above-ground ecosystem. Dotted across the landscape, termite mounds rose like miniature hill towns overlooking otherwise level terrain. It wasn't long before we noticed a curious pattern. The biggest trees, the ones to which the elephants and giraffes were drawn for shade and food, consistently grew on and close by active or abandoned termite mounds.

It turns out that termites have microbial helpers. The termites bring dead plant matter down into their mounds and feed it to fungi which break down plant tissues indigestible to the termites. And like any other gardener, the termites harvest their fungal crop and eat it.[2] In this interesting twist on agriculture, fungus-farming termites turn their termite mounds into nutrient oases that help support Africa's iconic wildlife.

Although we went to Africa, in part, to marvel at the impressive megafauna, by the end of the trip the pervasive, unheralded activity of termites captivated us as well. They built the foundation for the vegetation that supported the herbivores that, in turn, fed the carnivores that everyone thought were the main attraction.

Back home in Seattle our little garden reminded us of what we'd seen in Africa. Instead of fussing over and tending the plants on top of the soil, we should look beneath our feet—to the soil and the microbes and invertebrates that call it home. As we did, we came to think of the little-noticed and relatively unknown soil biota as a vast force of invisible allies working for us. After all, we saw firsthand that the secret to promoting life above ground was to promote life below ground. But we still had some things to learn about the microbial legions that support the health of animals—including ourselves.

7

TOO CLOSE TO HOME

I opened my eyes and looked around the room. Dave was sitting in a chair at the end of my hospital bed, engrossed in a book. Above his head, the hands of a clock stood at a few minutes before five. I turned my head and peered out the door at a bright, shiny floor and a busy nurse station. Relief flooded through my body. Five hours of surgery and the cancer that prompted it were behind me, I hoped.

I tried to sit up. F*&k! I sank back onto the pillow. Through the thin hospital gown I could make out the bumpy edges of a wide piece of cloth tape. I pulled the gown up around my waist and, beneath the tape, saw the outline of sturdy metal staples marching down my abdomen like a set of railroad tracks. I slid my finger over the first few that detoured around my belly button and kept counting. I stopped at number fifteen and closed my eyes.

Three weeks earlier, on a Friday in early May 2011, my doctor had left a message for me to call her about my recent Pap smear. Doctors never call to deliver good news, so I tried in vain all weekend not to think too hard about why she might want to speak to me.

"The lab pathologist made a special point of calling me. Your Pap smear results came back. The cells are very abnormal," Dr. Greaney said when I reached her on Monday morning.

I sat down at the kitchen table too stunned to say anything, twirling the phone cord in my fingers.

"The pathology report says carcinoma in situ."

"Carcinoma? That's cancer, isn't it?"

I like my doctor. She's frank, no bullshit, and always generous with information. But this time she didn't answer my question. My stomach fluttered and a wave of nausea rose up from my gut. I thought the lab results must be wrong, that there must have been some kind of mix-up. Those couldn't be my cells.

"You need to get a colposcopy as soon as possible."

"A what?" I heard "colonoscopy," which she'd reminded me about at my annual exam.

She sounded it out for me: "Colp-OS-ko-pee, a procedure that looks at your cervix under low magnification. You need to get one done."

That sounded bad, really bad, worse than a colonoscopy.

"*Right away,*" she added.

Her insistent tone left no doubt. There was no dodging this procedure, whatever it was called.

HPV is shorthand for "human papillomavirus," certain strains of which increase the risk of cervical cancer. Several years earlier I had discovered that I had one after a routine Pap smear. At the time, my doctor told me that up to 80 percent of all women carry one or more HPV strains linked to cervical cancer at some point in their lives.

I remember her cheerily adding that most people "clear the virus." She made it sound easy, like snapping my fingers. So, I figured my body would do that too, just clear the virus. After all, Dave was the one who got colds in our house, not me. I never got sick. So I didn't give HPV another thought—until now.

I made an appointment with a gynecological oncologist for a colposcopy and clung as tight as I could to the remote possibility that my Pap smear results had been mixed up with someone else's. Any hope I had of a mix-up evaporated when the oncologist told me he thought he'd seen a lesion lodged just inside my cervix. A week later the lab confirmed that the lesion was a small malignant tumor. The messed-up cells of the Pap smear belonged to me as sure as my blue eyes did. I fell apart. Dave, solid as ever, would hug and hold me, and I'd pull it together for a few hours or half a day, and then melt down all over again.

The malignant tumor needed to go, and I had two options: the first was chemotherapy and radiation, and the second was a radical hysterectomy. Normally, I like the word "radical." But paired with "hysterectomy," it meant I'd be losing my cervix and its partners, the uterus and ovaries.

Chemotherapy and radiation had their downsides as well, damaging other important parts of my body. I didn't like the idea of losing any of me, but reproductive parts versus brain, heart, and gut? I couldn't get by without the latter trio. Yet surgery entailed risk of infection or bleeding to death. Toxic chemicals or the knife? That colonoscopy was looking awfully good in comparison. In the end, I decided on surgery.

One side effect of this surgery was guaranteed—menopause. I had less than a month to study up on it. I consulted books and far too many websites. How could such a normal biological process be as bad as everyone made it sound? Nearly everything I read likened it to a Dantean descent into a hell of madness and health problems. And that was if you were lucky enough to enter menopause at a natural pace. I was leaping off the hormonal cliff.

The ability to bear children makes a woman's body both wonderfully, and not-so-wonderfully, complex. And all this complexity creates opportunities for the marketing forces of consumer products and modern medicine. Now, facing an abrupt menopause, a truly stunning array of drugs, supplements, and hormonal fixes were enticingly dangled in front of me, promising to address hot flashes, depression, and myriad other physiological changes.

But it wasn't so simple. Each cure seemed to be surrounded by a plethora of caveats and conflicting studies. My bones would start dumping calcium once my estrogen levels plummeted, and so I should take calcium supplements. And yet, there are different forms of calcium, so I should only take the type that is easily absorbed. But I shouldn't take too much of it because it would clog up my arteries. But, if I took hormones instead, I wouldn't need to take calcium. And yet, a troubling set of studies implied that I shouldn't take hormones for too long because I'd set myself up for more problems down the road, among them an increased risk of breast cancer and stroke. Or maybe not. Other studies indicated that these studies needlessly alarmed newly postmenopausal women because the test subjects were older postmenopausal women. Of course, any mention of cancer only heightened my anxiety. It didn't take long before I felt tears rolling down my cheeks again, and I escaped into the garden.

I'd be laid up for at least eight weeks after surgery and couldn't fathom missing the chance to garden over the summer. So I flew into a frenzy for several weeks before the fateful day. I set twine and supports for some *Clematis* vines I was training to grow up into several Hinoki cypresses and wished them luck. I wrestled my living fence into shape with a summer's worth of pruning cuts and told it I was sure it would recover. My collection of potted trees and shrubs that ringed the patio needed their annual grooming, so I frantically pruned, repotted, and mulched each of them. Despite my best efforts to try and lose myself in a whirlwind of gardening, I spent nearly every moment fretting about what lay ahead of me.

The days ticked by, and on June 6 I set my pruning shears to the locked position, slipped them into their leather holder hanging on my garden cart, and wheeled the cart into the garage. I tidied up the shovels and pitchforks and rakes and pulled the rickety garage door closed. Though already well past ten in the evening, I walked around one last time, watching the sun's lingering glow on the horizon slowly give way to an inky royal-blue night sky over my head. The garden looked so gorgeous in the twilight. The air was perfectly still and quiet. But I felt none of that peace. I sank to my knees on the soft earth as another torrent of tears erupted from my tired eyes. Hours later and still dark, Dave and I left for the hospital.

Two weeks later the surgeon went over the pathology report with us. There were clean margins on the excised tumor and no sign that the cancer had spread elsewhere. My malignant tumor was history. I was thrilled and incredibly grateful.

Despite my good prognosis, I still felt rattled and afraid. I thought about my friends and family who'd had cancer—my aunt, Dave's mom, and two members of my book club. My own mother had been diagnosed with melanoma when she was fifty-two and died just two years later.

Cancer scared the hell out of me and fascinated me at the same time. I wanted to learn as much as I could about it. So I resumed my research postsurgery. I discovered that my particular variety, cervical cancer, is the fourth-most-common cause of cancer death among women around the world. In the United States it drops to fourteenth place because of widespread screening and the kind of aggressive treatment I was fortunate enough to receive. I also learned that in 1976, a German virologist, Harald zur Hausen, proposed that HPV played an important role in cervical cancer. Although his hypothesis was met with initial skepticism, he was able to definitively link HPV to cervical cancer in the 1980s, a discovery that earned him a Nobel Prize in 2008. Zur Hausen's discovery opened the door to an HPV vaccine, introduced in 2006 for people under the age of twenty-six.

Like most viruses, HPV comes in different subtypes, two of which (subtypes 16 and 18) are linked to over half of cervical cancers. But, HPV is proving wily. In 2013, researchers from Duke University and the University of North Carolina discovered that in addition to HPV subtypes 16 and 18, about half a dozen other subtypes of HPV are linked to cervical cancer in African-American women.

Another interesting finding links HPV to cancer in the head and neck. A 2011 study published in the *Journal of Clinical Oncology* found a significant increase in the number of oropharyngeal (nose, mouth, and throat) cancers related to HPV in the United States. The proportion of people with this type of cancer increased from 16 percent in the late 1980s to 72 percent by the early 2000s. Curiously, the sharpest increases in these cancers are among otherwise healthy nonsmoking white men in their thirties and forties. The study concluded that if the trend continued, by 2020 the number of HPV-caused oral cancers would exceed cases of cervical cancer. Another study in 2013 found similar trends in other economically developed countries. Some researchers think that HPV may also play a role in cancers found in other types of mucosal tissue, including the lungs and esophagus.

Despite what is known about the cancer-associated strains of HPV, researchers still don't fully understand all the steps between an HPV infection and full-blown cancer. According to another 2013 study, it appears that, in part, the way the virus slips its genes into the DNA of a host cell plays a role. Apparently, HPV's genes arrive with hurricane force, knocking nearby genes topsy-turvy. Some become rearranged. Others are eliminated. Trouble can follow if the disrupted genes play a role in suppressing tumors, or if the disturbance activates other genes with the potential to promote cancer.

Although HPV was my personal nemesis, other microbes can also cause cancer. The bacterium *Helicobacter pylori* is linked to stomach cancer, and the hepatitis B and C viruses can lead to liver cancer. Along with HPV, these three microbes are responsible for almost one in five cancers worldwide.

While cancer-causing microbes were a surprise to me, it wasn't a leap to learn that just over a third of all cancers are thought to be linked to diet, including breast cancer in postmenopausal women, colon cancer, and prostate cancer. The diet–cancer connection resonated strongly with me. I'd long believed that you are what you eat, but that didn't mean I always ate well.

Like most kids, I lobbied my parents for what to buy at the store and sometimes got my way. I loved pastel-colored Froot Loops adorned with chunky sugar crystals, and golden mininuggets of Cap'n Crunch. Almost as soon as these cereals made it inside our house, they disappeared. My brothers and I inhaled them for breakfast. After school we raided the cupboard and stood in the kitchen eating handfuls of the crunchy sweet cereal straight out of the box.

Growing up in Littleton, Colorado, there was a Pizza Hut just up the hill from my middle school. My friends and I loved that place. Pitchers of root beer were half price as long as you ordered them at the same time as the pizza. And the dark interior and red faux leather semicircular booths transported us to a world where we felt grown-up and miles from the teachers and classrooms a stone's throw away.

A few years into high school, the way I thought about food began to change. The father of one of my best friends made an annual feast of guacamole and enchiladas for his graduate students. I had always watched my mom or dad cook, but this grand event was in another league. One year I was recruited to help with preparations. Peeling and mashing nearly two dozen avocados, chopping what seemed like endless amounts of purple onion and garlic bulbs was on a scale I had never experienced. Mixing all the fresh ingredients together verged on hypnotic. A heaping handful of dark red chile powder swirled into a mountain of bright green avocado. The juice of numerous limes loosened up the mix, the scent of citrus wafting out of the bowl. Cumin, salt, and oregano added more flavor, and then in went the onions, garlic, and a touch of finely chopped tomato and cucumber. After that experience, my friend and I occasionally took over her kitchen. We made enchilada sauce from scratch and experimented with creating different fillings. By the time I started college, I loved to cook. My housemates and I ate together, each of us responsible for cooking dinner one night a week. We all tried to outdo one another. I'd soon exhausted the only two cookbooks we had, Molly Katzen's *Moosewood Cookbook* and *The Enchanted Broccoli Forest*, and began inventing new recipes.

Years later, though several years before my cancer diagnosis, I made a career move to the public health field. The Seattle–King County Health Department was looking for someone with experience in environmental planning and a policy background to work in a new program they called "the built environment." Public health departments across the country had borrowed this term from the fields of urban design and landscape architecture. The idea is that the way cities are laid out influences the choices and opportunities that underlie a person's health and well-being.

I found this idea intriguing. Although I didn't know much about public health, I knew a lot about the effects of the natural environment on plants and animals. I'd worked as a field biologist in California after college, surveying and inventorying endangered species and the places they lived. When Dave and I moved to Seattle, wild salmon, and the rivers and streams they inhabit, became a part of both of our careers.

For over a decade I worked along the lower Cedar River, a dozen miles southeast of Seattle. With my ecologist and engineer colleagues I arranged voluntary buyouts of frequently flooded homes so that we could remove levees and replant native shrubs and trees along the river. This made way for the river to once again roam across its floodplain. Our philosophy was to restore the natural processes that sustain salmon and let the river do the heavy lifting of recapturing spawning gravels and scouring out deep pools. Indeed, almost immediately after our interventions salmon would begin to use the new habitat.

I took all my experience from studying river systems and salmon habitats and applied it to my new job. Interestingly, my public health colleagues even used a term familiar to me. To them an "upstream" approach meant treating the root causes of disease and poor health. Instead of looking at the quality of a floodplain or the conditions in a watershed, I began to consider other factors—the quality of air in a town or city, the level of crime and violence in a neighborhood, and even the number of parks and the presence of sidewalks and grocery stores near people's homes. I was surprised to find that, when averaged across the populace, these and other upstream factors typically influence a person's health more than genes or visits to a doctor's office. It stood to reason then, if habitat restoration could improve the health of salmon populations, as I'd seen on the Cedar River and its tributaries, why couldn't the same approach work for people and the places they lived?

From one project meeting to the next, new and inescapable facts came to my attention. Nationally, about one in three kids is obese or overweight, many of whom go on to develop chronic health problems like type 2 diabetes and heart disease. Even without a public health background, no one had to tell me that these conditions are the life-sapping ailments of parents and grandparents, not children. Epidemiologists talked of chronic disease rates in the future putting a downward kink in the otherwise upward-trending longevity of Americans. Unless something changes, today's youth won't live as long, on average, as their parents' generation.

For the most part, I considered myself beyond the reach of such statistics, until, that is, I faced cancer. At the time of my diagnosis I began to think much more about my health. Removing the centimeter-sized tumor was an obvious necessity, but it seemed like a partial recipe for well-being. What about the rest of my body?

I cast my net far and wide in search of answers. I consulted with friends, did research, followed up on referrals, and began seeing Heidi, a naturopathic oncologist. I had no experience with naturopathic medicine and wasn't sure what to expect. I had my first appointment with her a few days before surgery. She asked me questions, lots and lots of questions. In fact, in our hour-long appointment, I did most of the talking. Part friendly interrogator and part health confidante, her warm brown eyes glanced up from her notepad to meet mine with each new question. She reassuringly smiled with an "Uh-huh, tell me more about that." As we talked, I began to see a pattern in her line of questioning. She zeroed in on three themes: What did I eat? What was the stress level in my life?

What did I do for exercise? No doctor had ever so bluntly or thoroughly talked with me about these things.

I described my standard breakfast, a tall latte, and if the coffeehouse had them, a scone from Macrina or Essential, my favorite local bakeries. I told Heidi that eating breakfast at home took too long. Plus, I was groggy in the morning and usually had to rush out of the house to catch my bus to work. By early afternoon, the caffeine and scone wore off, and I was ravenous. My lunch was usually leftovers. But, late lunch led to late dinner, which I realized in retrospect, contributed to not being hungry for breakfast.

She hinted that I'd be better off with regular mealtimes, and no daily scones, coffee, or wine. I was speechless and chagrined. After a moment I pushed back. How, exactly, do I live in Seattle and not drink coffee? How do I not go to Macrina or Essential, or take a break from work in the afternoon and go to Caffe Vita? And, wine? Wine was part of dinner. Wine got left off the food pyramid, or whatever it's now called. How do I not drink wine?

Heidi patiently pointed out that I shouldn't have these things *all* the time. The caffeine, she told me, interfered with my natural appetite. And my body saw the white flour of a scone as sugar, and wine as more sugar. With deceptively simple cheerfulness she added that there were lots of other things to eat! Eating a real breakfast and cutting back on coffee didn't mean I could never have coffee or scones again. Wine shouldn't be taboo, but it shouldn't be everyday. My habits needed to become occasional things.

By this point, I was getting somewhat better at accepting things I didn't want to hear. But still, I looked for wiggle room, a way not to change. I proposed a latte every other day instead of every day. Heidi masterfully dodged the "yes" I wanted, and the "no" I didn't, by encouraging me to see if I could go without for longer stretches of time.

Heidi also wanted to know what Dave and I ate for dinner. I could see, even as my answer tumbled out, that telling her vegetables made it onto the dinner plate most nights wasn't good enough. She flipped to a clean page in her notebook and drew a circle. Then she drew a vertical line from the top of the circle to the bottom. I was supposed to fill half the plate with plant foods, especially cruciferous greens, other vegetables, and fruit. She drew another line coming off the line dividing the circle in half. Now the plate looked like a half peace sign. Unprocessed whole grains were to occupy the smallest slice of my dinner plate. The third remaining area should hold protein from plants, like beans or legumes. Animal protein sources were okay, just not too much. And whatever I put on my plate should not be tumbling off the edges. She also told me that one of the best strategies for quelling hot flashes was to change what and how I ate. Less caffeine, alcohol, and sugar, she told me, could help a lot. I wondered how I could possibly make all these changes.

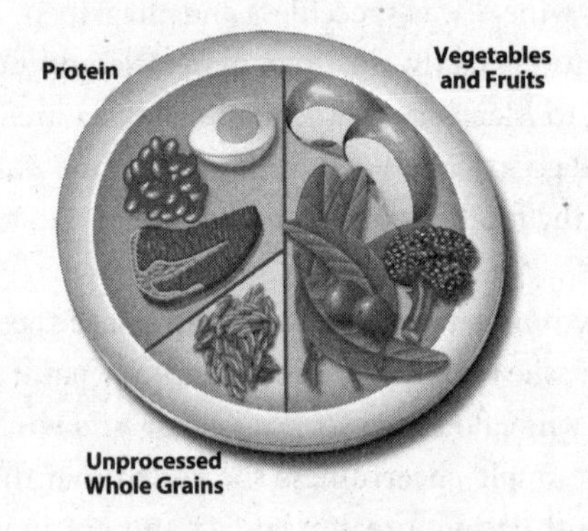

Heidi's Plate. Our starting point for rethinking meals to promote overall health, support the immune system, and help prevent cancer.

On one point Heidi was adamant: that I eat three proper meals every day. Three meals a day? I hadn't done that for a long, long time. First, a proper breakfast would take time, and I had no time in the morning. Second, I saw myself ballooning to the size of a cow if I ate three solid meals a day. Pushing back, I put a question to her, "Why three meals?"

Heidi told me that the way I ate—erratic mealtimes—turned my blood sugar into a pinball. She gently explained that when my blood sugar surged and dove it stressed my organs and cells, making them work harder to keep glucose at a steady level in my body. When blood sugar climbs and plummets over and over, it causes inflammation. And eating sugar straight up does the same thing. Inflammation? Swollen lymph glands from my childhood were all that came to mind. But what really got me was the next thing she said—inflammation helps fuel cancer. *That* riveted my attention.

I pictured her admonitions, billboard-sized, looming above me, and craned my neck to take everything in. No one had ever told me these things before. I felt as rattled after my conversation with Heidi as I had when I'd gotten off the phone with Dr. Greaney.

I knew that diet and inflammation weren't the proximal cause for my type of cancer, but if unnecessary inflammation could fuel cancer and other maladies, I didn't want it. One bout of cancer was enough. I asked her for the piece of paper with the plate sketched on it. Somehow I would change.

Over the course of my appointments with Heidi, I recognized a duality in food I had never seen before. I assumed that living in Seattle meant I didn't need to think too much about what I ate. It seems everything is healthy here. The coffee is organic. The flour is organic. The pasta is whole wheat *and* organic, gluten-free if you prefer. Plain sugar is out, cane sugar is in. Cows eat grass grown on pure rain. Chickens walk, cluck, and then pluck fat grubs from verdant ground. Seattle's phenomenal bakers and butchers and their cornucopia of creations—sweet baked goods, savory breads, cheeses hard or soft, and every cut of meat imaginable—had reeled me in again and again.

Today, all across America there is an astonishing array of food choices for every palate and pocketbook level. We can choose foods grown close to home or halfway across the world. We can shop at Whole Foods or a discount grocery outlet. We can visit farmers markets on the weekends or every day of the week depending on where we live. We can slave in our kitchens, fetch takeout, flock to food trucks, or eat from convenience-store fare. We can have fast food, fine food, or opt for something in between. We can enjoy cuisines from around the world sometimes blocks from our front door. While the picture is certainly different for the almost 50 million Americans who face hunger or food insecurity, most of us are drowning in a sea of deliciousness.

I realized that no matter where on the food spectrum one eats—or where on the socioeconomic ladder one falls—most of us naturally make a beeline for sugars and fats, especially the former. A long time ago, before we lived in cities or cooked in kitchens, before agriculture even existed, such inherent preferences gave us a leg up. Sugar was relatively rare other than in fruits. And seasonality further limited our consumption. So great was our desire and need for quick sources of energy, we would risk life and limb to steal honey from a beehive-crowned tree. Industrializing fields and factories transformed a once-rare food source that helped us survive into something as seductive as opium.

Heidi helped me realize that I had my food proportions all wrong. Not the amount as much as the mix—too few vegetables and fruits and too much of other things.

And the irony of it all? I actually like fruits and vegetables! But my behavioral software keeps humming away in the background, shaping my actions. Following preferences honed through evolution, certain foods finagled their way into my mouth and burrowed deep into my psyche. Grain-based carbohydrates refined into simpler sugars light up the pleasure centers of my brain and make me smile. Who doesn't love a properly made and toasted bagel? All the better slathered with cream cheese or butter and a nice jam. Add some coffee and one can go, and go some more, as I had long done.

My palate is not hardwired for mineral- and phytochemical-laden plants, especially leafy greens. Neither is Dave's. These substances imbue vegetables (and some fruits) with astringency and bitterness. *Blech.* Next to sweetness and sumptuous fats, why would anyone naturally be attracted to these foods?

Vegetables and some fruits suffer the same problem as Labrador retrievers at the Westminster Dog Show who, since its inception, have never won "Best in Show." You just can't doll up a Lab the way you can a poodle. Their ears always hang down, their fur is too short to curl or gather together and adorn with a bow. Their appearance is stubborn and unyielding.

Unlike meats, dairy, and grains, vegetables and fruits don't lend themselves as readily to transformative preparations. It takes some skill and imagination to gastronomically seduce people with turnips, cabbage, or pinto beans. Fruits, with their naturally high sugar level, are an easier sell, but even they require cover. People do not *really* eat a piece of pie for the fruit, all too often a mushy clump of coagulated cellulose. Crust and sugar are the point of pie. And a Danish roll? What fool eats the fruit puddle in the middle and tosses the pastry? That apple-pie-like thing

McDonald's makes? Kids don't browbeat their parents for them because they love apples.

What it comes down to is that the hallmarks of humanity and civilization are foods we can cultivate in large quantities, store for later use, and process into pleasing creations we crave. Bakeries, food trucks, and restaurants don't specialize in selling apples, kale, or cantaloupes. For the most part, they market, prepare, and sell dishes and meals built around milled grains, meats, and dairy.

Vegetables and fruits are holdovers, staples from another time. Early on in human evolution these mineral- and phytochemical-laden gems dominated our diet. We know this because our prehuman ancestors sported a giant sagittal crest at the top of their skulls. Massive muscles draped off this cranial ridge and secured their jaws. These muscles enabled them to eat from nature's table—fibrous leaves, thick roots, and the leather-like skin of fruits filled with dry pulp. They spent up to six hours every day chewing, grinding, and swallowing plant matter. And this was after they'd spent a good part of the day grubbing around in the dirt for edible roots, or searching for fruits hidden amid branches and leaves. Finding and then chewing food was a full-time job.

We did much the same as hunter-gatherers. Then we found an escape from this jaw-numbing drudgery and endless search for sustenance. We used our big brains to catch animals, tame fire (and some animals), plow fields, and grow storable food packed with calories. And the icing on the cake: we could trade or sell our wonderful, ingenious creations—bread, candy bars, or beef jerky—to each other. Then we sat down and ate a lot of such things, except, of course, during famines.

Navigating food choices in the modern world with my ancestral circuitry led me to a conundrum. I had routinely passed over the foods that were best for my long-term health. Most perplexing and disconcerting of all was this: how and what I ate was out of sync with what I believed. I thought food and eating should nourish my body and mind, not lead me down a path of illness and disease.

What I had learned in the months after surgery, on my own and from Heidi, led to overhauling my diet, which, since I was chief cook, also meant overhauling Dave's. He was not thrilled. I could relate. Just a short time ago, I hadn't been so thrilled with it either. I had been annoyed, irritated, and defensive. But the way Heidi framed eating persuaded me. Nothing was completely off-limits. It was just that my daily consumption of caffeine, alcohol, and refined carbohydrates and sugars would need to change.

I shared my new edict with Dave, cheerfully announcing that we'd be eating more greens, beans, and fruits, and less refined grains, meat, and dairy. My proclamations garnered head nods and "uh-huhs" but no solid endorsement. I told him about the sacrifices I'd be making. How I'd wake up earlier so I'd have time to eat a real breakfast. That I'd cut back on afternoon sugar bombs, indie bakery visits, lattes, and wine. *Waaay back.*

When you say something out loud, it becomes more real, even if it sounds outrageous. I couldn't believe I was actually telling Dave that I'd be kicking my coffeehouse habit. He didn't believe me at first. Yet, slowly but surely, everyday habits became occasional delights. Little by little, I reeled him in, and he (mostly) accepted a new way of eating too.

I am not a glass-half-full person. I never would have thought that anything good could come from cancer. In the beginning, my diagnosis felt like a death-around-the-corner sentence. I had no idea how to combat feeling hopeless and helpless. But I began to see that there really were things within my power, things I could do, to change my health. I'd chart my own course through the sea of deliciousness instead of drowning in it.

Ask a true gardener what to do in almost any predicament and odds are they'll turn to plants. We'll use flowers to mend a relationship or cheer ourselves up, and plant trees to change our view. So I cooked up a grand plan for how to put the long-ago and good-for-me foods back at the center of my plate. I would grow them.

It seemed a natural thing to do. We had just put in vegetable beds a few summers before. Initially, I picked plants whose looks I liked and that did well in Seattle's modestly warm summers. I found a bean called Roma whose meaty pods rode the vines into the sky toward the finish line of the sun. The size of my basil leaves astounded even me. They had the lushness of an ornamental plant. A single leaf covered the palm of my hand.

We scarfed the beans and basil, but I refused to harvest the onions. They looked marvelous against the seaweed-green backdrop of new growth on the Hinoki cypress. Their orbs of tiny flowers tinged in pink floated like fairies above the beds. Round, glaucous leaves came to perfect points and swayed in the slightest of breezes. The flower clusters drew a crowd of bees, parasitoid wasps, and hoverflies. These plants so entranced me that I spared them the kitchen knife.

But these vegetable beds were in my past, before cancer. From now on, no more clemency—we would eat all the vegetables I grew. And I switched over to mostly growing plants in the Cruciferaceae, or mustard, family. Chock-full of cancer-fighting phytochemicals, they include rugose-leafed dinosaur kale, Russian red kale, purple kale, curly kale, bok choy, and broccoli.

The garden had nourished my mind; now I'd use part of it to grow an edible pharmacopeia straight out of the earth. To help it help us, I'd dedicate my precious worm compost to the veggie beds. It would grow more soil life to kick the beds into high gear and help load up the vegetables with all-important minerals, vitamins, and phytochemicals. I knew that once plucked from the soil, vegetables immediately start to lose their nutrient value. Of course, the biologist in me knew they'd died and begun their journey back to the earth, making the mere ten steps between kitchen and garden even more significant. Before my cancer diagnosis, I infrequently shopped for veggies at the Saturday morning University District Farmers Market. Now I became a regular, using it to supplement what I grew at home. The greens and other vegetables I purchased there might not be from my garden, but they were almost as fresh.

Plant foods began to occupy much of the space in our refrigerator, and with Heidi's plate in mind I began experimenting. Greens of one sort or another served as the central ingredient, the way I used wood chips as the base in my mulch recipes. I pictured the plate Heidi drew for me and added a protein of some sort and a dash of whole grains. Vegetables, like onions and garlic, and a fat, like olive oil or coconut oil, laid the foundation of flavor. And just like when I made my mulches, I combined what I had around with what struck me in the moment. I used herbs, spices, condiments, and more vegetables to take lunches and dinners in different directions—Asian, West African, Mediterranean, Middle Eastern, or American Midwestern. Heidi's words echoed in my head. There *were* lots of other things to eat.

Eating to prevent cancer netted me more benefits. I lost the extra pounds I'd been lugging around for a decade and reversed my upward-creeping blood pressure. My base metabolic panel, cholesterol, and lipids dropped into the healthy range.

The surgically induced hot flashes were as advertised—raucous! But, changing my once-everyday fare—refined carbohydrates, lattes, and wine—to occasional fare tamed them to a whisper. When I added physical activity like yoga, walking, and the hauling and digging that comes with gardening, my symptoms went to near-silent about a year after surgery. Changing what and how I ate delivered the most elixir-like effects to my body and mind I have ever experienced.

In the months after surgery, I was more aware of my health than ever before. I tried not to let cancer dominate my thoughts, but it was a constant battle. I had quarterly checkups with the oncologist. At my first appointment, a nurse walked me down the long hall leading to the exam rooms. Midway down the hall there was a room with oversized windows featuring a stunning view of Mt. Rainier with Lake Washington in the foreground. A half-dozen women of all ages, some with friends and others by themselves, were reclining in big La-Z-Boy–style chairs, soaking up the sun and scenery. Except for the intravenous drip-bag stands by their sides, the scene reminded me of the relaxed, carefree way Europeans lounge about at ski resorts. But some of these women had no hair and pale, drawn faces, while others appeared completely healthy. They were all receiving chemotherapy. I could very easily have been one of them, staring out that window and wishing I was somewhere, anywhere, else.

Sometimes I could peek in that doorway and remain composed. On other visits, my anxiety rose. I imagined different scenarios, none of them positive. What if the margins on the tumor weren't really as good as the surgeon thought? What if the cancer had traveled beyond the tumor? What if it was brewing somewhere deep inside me?

These what-if scenarios, as unpleasant as they were, kept me on course. Heidi applauded the significant changes I had made to my diet, while emphasizing the importance of maintaining a strong immune system and minimizing excessive inflammation. I'd already learned quite a bit about the impact of food on health, but I knew woefully little about immunity and inflammation—and next to nothing about the relationship between them. There was nothing I could see, or touch, or smell, no clue indicating what direction I should go in specifically to boost my immune system while keeping inflammation at bay. And all along was the reminder that my immune system had already failed me on two counts. It hadn't ousted the HPV or the cancer.

I came upon the first bread crumb almost one year to the day after my surgery. Dave and I were on a short vacation in the Bay Area and had fled the bustle of San Francisco for the serenity of Muir Woods north of the city. Late in the afternoon we stopped at the Pelican Inn, a thatched-roof, English-style pub nestled among coastal dunes in rural Marin County. We settled in for a cup of tea when I spied a headline in a copy of the *Marin Independent Journal* that had been left on our table. The article mentioned viruses and the human microbiome project.

Viruses?

That word will forever catch my attention after what I'd been through with HPV. I picked up the paper and read the first paragraph. Several hundred scientists had just released census-like results about the raft of microbes that eat, excrete, reproduce, and die on and in the human body. Other life forms eating and excreting on us and in us? It sounded creepy and intriguing, just the kind of biological notion that lodged in my gardener brain and started to itch.

Then, upon returning home, we found a letter from my father in the pile of mail that had accumulated while we were gone. A retired aerospace engineer turned landscape painter, he regularly sends me articles he cuts out of newspapers. This one, from the *Denver Post*, trumpeted how bacteria had a good side and played key roles in human health.

I was vaguely aware of the Human Microbiome Project that the newspaper articles described, but I didn't know that the scientists running the project had published their findings. So I read the debut set of papers in *Nature* and *Science* and then dug up some of the papers listed in those papers, and then more papers listed in those papers.

I was agog. Bacteria, viruses, and more were at work in my body and Dave's body in ways I never imagined. Not everything was an HPV-like menace, far from it. It turns out that immunity is partly about the quality of the environment in which microbes live—the lay of the land that is my body and every body. Human microbiome scientists spoke of an incessant dialogue between microbes and their hosts. Molecules were the words of their common language. And this conversation wasn't just idle chatter. Microbes, especially those conducting their business in the gut, could shift a person's immune response, the level of inflammation, and other aspects of their physiology every which way.

I waded into this new world of microbes within us, pulling Dave along with me. It took both of us to decipher the near-impenetrable cell and molecular biology jargon masking the paradigm-busting ideas of microbiome researchers. What we read echoed Lynn Margulis's ideas and thinking from over three decades ago. The human body is one vast ecosystem. Actually, it's more like an entire planet with a rich palette of ecosystems, as different as the Serengeti and Siberia, each hosting multitudes of microbes. Paper after paper left us stunned.

From the perspective of a microbe, I am a durable, living trellis—inside and out—on which vast numbers of those microbes cling, climb, and grow. For every one of my cells, I harbor at least three bacterial cells.[1] They live all over and in me—my skin, lungs, vagina, toes, elbows, ears, eyes, and gut. I am their homeland.

I am not who I thought I was. And neither are you. We are all a collection of ecosystems for other creatures. But it's not just the microbes themselves that add to who we are. They add to our genetic repertoire. Bacteria alone bring about 2 million genes into our bodies, several hundredfold more than the roughly 20,000 protein-coding genes of the human genome. Add the genomes of other members of our microbiome—viruses, archaea, and fungi—and the number of microbial genes in our bodies could be as high as 6 million. Most of the time this is a good thing. Their genes make it possible for us to assimilate dozens of essential nutrients critical to the health of our immune, digestive, and nervous systems.

A little taxonomy is helpful to fully grasp the diversity of the human microbiome. Recall that a phylum is the third grouping of life, just beneath kingdom. To date there are about fifty known phyla of bacteria. Across all humankind, our intestines alone host species from up to twelve of those phyla. For comparison, all the plants on the planet fit into twelve phyla, and the vast majority of animals belong to just nine. If you are not yet beginning to feel uncomfortable or insignificant, we and all other vertebrates—fish, amphibians, reptiles, birds, and mammals—tidily fit into just one phylum, the Chordata.

Human microbiome researchers point to two phyla of bacteria in particular that dominate the human gut—Bacteroidetes and Firmicutes. Of the other ten phyla composing our gut microbiome, bacterial representatives commonly hail from the Proteobacteria, Actinobacteria, and Verrucomicrobia if you are a Westerner living in a developed country. But if you come from a hunter-gatherer group, or a rural village in Africa or South America, your gut microbiome composition is quite different and far more diverse.

Drawing on a cast from at least twelve different phyla means that none of us has the same mix and relative abundance of species. Despite the difficulty of making the species concept stick when it comes to bacteria, it's estimated that the collective human gut microbiome of Westerners contains in the neighborhood of about 1,000 bacterial species and many different strains of these bacteria.

Although microbiome research to date has focused almost exclusively on bacteria, add the other types of microbes and the human microbiome spans domains and kingdoms! Add another player, the viruses that get no perch on the tree of life, and it is easy to see the extraordinary chimera that we really are.

Of all our bodily habitats, the richest in terms of abundance and diversity is our twenty-two-foot-long digestive tract. In particular, the last five feet—our colon—houses almost three-quarters of our gut microbiome, many *trillions* of denizens. Who would suspect that the microscopic inhabitants of the lowermost part of the gut rival the visible biodiversity of Earth itself?

More astonishing, the vast majority of microbes that live in our gut have never been cultured. They cannot survive outside of our body. And so until very recently, we had little idea of the species and strains composing not only the gut microbiota, but the entire human microbiome. Gene-sequencing technology, however, grants us a new window onto the microbial world.

I was also surprised to learn that about 80 percent of the immune system is associated with the gut, in particular the colon. Immunologists have a not-so-fancy name for the lion's share of the immune system—"gut-associated lymphoid tissue," or GALT. Here was another key factor underpinning my health that I hadn't considered—the microbes in my colon. I found this astounding.

The more I learned about it, the more I pictured my GALT as a biological version of Harry Potter's sorting hat. Remember how unusual that hat was? It twirled and hopped its way through the ranks of students entering Hogwarts each year and analyzed their character to assign them to the proper house. The sorting hat had power, an all-knowingness about who was up to what.

Change the shape of the sorting hat to a tube and imagine slipping it on where the stomach transitions to the small intestine, pulling it all the way down to your bottom. While the sorting hat's power came from magic, immune cells and tissues imbue GALT with its power. Immune cells even make up part of the colon wall itself.

I began to see why my gut is so central to my immune system. On one level my mouth is a gateway—an organic hybrid, part Grand Central Station and part Ellis Island. My tube-like gut holds the vastness of the outside world inside of me so that my immune system can come to it. My GALT must sift through everything I eat and drink because I may have ingested a pathogen. Any breaches in my gut and I'll be in big trouble. If the outside world—whether pathogens or partially digested food—circulates willy-nilly around my body, my immune system would go into overdrive and either burn up or burn out.

The colon wall is scarily thin, only one cell thick. Thankfully, a couple of things backstop colon cells. Like a thick layer of protective soil, mucus sits on top of the cells facing the lumen, the name of the space inside the colon. Lying against the outside of the colon wall is a special compression-sock-like membrane that holds the GALT snug against the biggest potential source of pathogens in the body.

Colon cells churn out mucus, a lot of it, constantly. Mucus serves multiple purposes. It keeps the contents of the colon rolling along and forms a shield that prevents bacteria from getting too close to the GALT. And, when needed, colon cells add compounds to mucus that deter pathogens from attaching to the colon wall. Chemically, mucus is a carbohydrate. Like the exudates that roots release into the rhizosphere, mucus is also a food source for bacteria.

To get a sense of this world, imagine shrinking yourself to the height of a millimeter or so, throwing on a wetsuit, and climbing inside the colon. (We'll pretend there isn't much in the lumen at the moment.) At your feet and overhead, you'd see the mucus-covered single layer of cells composing the colon wall. Pull out the microshovel you brought along and dig into the ground beneath your feet.

After a few shovelfuls, you should be able to see the entrance of a pouch-like feature. This is the top of a crypt, each one formed by a tiny infolding of the colon wall. A crypt is narrower at the top and wider at the bottom. The luminal surface of the colon is peppered with thousands of crypts. New colon cells develop in the bottom of the crypts and migrate upward to the luminal surface in conveyor-belt fashion. Though you're microsized, you won't fall into the crypt. The opening is far smaller than your millimeter-tall body.

Perhaps the strangest thing I learned from my microbiome research is that in experiments with mice, researchers found an abundance of certain bacteria living deep down inside the crypts of the colon, an eddy-like niche out of the colonic fray. They too found this strange. Both the mucus and the crypts were long thought to be inhospitable because of antimicrobial compounds that colon cells secrete.

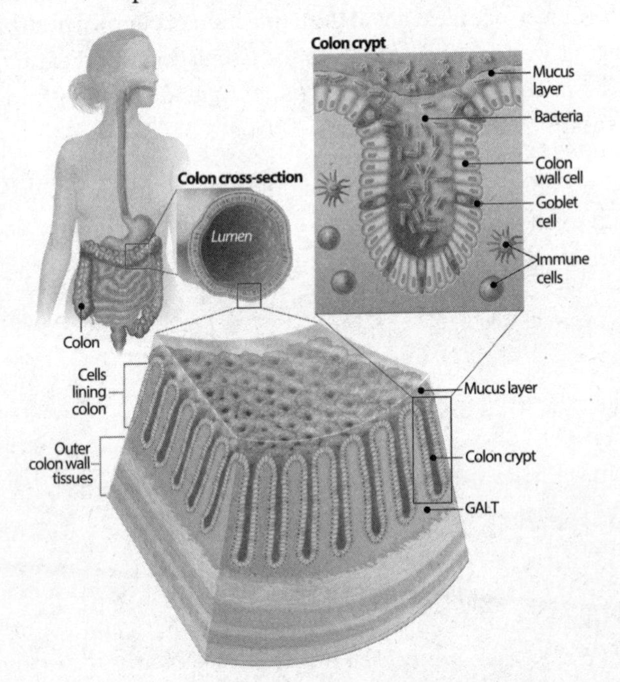

Life on the Inside. More of our microbiome lives in our colon than anywhere else. Most bacteria inhabit the lumen, although certain bacteria reside in the crypts of the colon or in the layer of mucus lining the colon wall. Immune cells congregate in the GALT (gut-associated lymphoid tissue).

But life in a colonic crypt is a good one for bacteria that find ways to dodge the antimicrobial compounds in mucus. For them, the crypts and the mucus coating the colon wall offer protection from predators, as well as food. A special type of cell found along the sides and bottom of each crypt, known as goblet cells, couldn't have a more apt name. These cells exude carbohydrate-rich mucus that crypt-dwelling bacteria can lap up. That some bacteria not only penetrate the mucus, but also live peacefully alongside colon cells and the GALT, is the kind of discovery that is leading to new ideas about microbes and their influence on our health and well-being.

As Dave and I were swept up in learning about the microbiome, we realized that much of the latest research on this subject centered on immunity and inflammation. And so, just as we had unlearned conventional ideas about soil fertility when we discovered the world of microbes beneath our feet, our ideas about the foundation of our own health began to change as we dipped our toes into the fascinating world of immunology. It seems that our moms were kind of right all along—it's who you are on the inside that really matters.

8

INNER NATURE

Loki is a small black Labrador retriever whose insatiable, indiscriminant appetite got him in some big trouble as we were writing this book. Normally, he meets the world with wild curiosity, but for several days he had been lethargic and breathing with seal-like snorts. Then one morning, in the middle of a brisk walk, he suddenly stopped, stubbornly planting his feet. A moment later, he was looking forlornly at the sidewalk, his just-eaten breakfast pooling around his feet. Before he could wolf it down a second time, we turned around and headed home. On the way back his breathing became even stranger, big deep breaths alternating with gasps.

By the end of the day Loki landed in critical care, where he was diagnosed with pyothorax—pus in the chest. According to the vet, a particle that Loki had either inhaled or eaten had somehow escaped his lungs or esophagus and worked its way into his chest cavity. But the problem wasn't so much the particle itself as the bacteria on the particle. They piggybacked their way into Loki's body, made themselves at home, and began to reproduce. One became two became four and so on. In less than a week, the original bacteria had spawned billions of copies of themselves.

Inflammation was how Loki's immune system dealt with the invading horde. Normally, inflammation is one of the body's most helpful and necessary biological processes because it's central to killing pathogens and healing from the problems they cause. Loki's fever and sky-high counts of immune cells were among the signs that told the vet his immune system was working. But in Loki's case the pyothorax developed into an over-the-top inflammatory response that threatened to kill him.

Pus, lymph, and immune cells associated with inflammation had accumulated in his chest cavity and were slowly crushing his lungs. A grave-faced vet presented us with the options. We could euthanize Loki before he suffocated, or move quickly to try and treat him. But even with treatment, Loki's chances were slim. Many dogs die, the vet warned, from septic shock due to the runaway train of inflammation brought on by advanced pyothorax.

Loki came to us from Guide Dogs for the Blind, where for a short time he was destined for a life of service. However, his boundless joy extends far beyond sidewalks and stoplights. When he sees other dogs, he launches into jubilant, bloodhound-like barking before barreling straight for them. The guide dog trainers knew that this trait (among others) would lead to mayhem, and they relieved Loki of his would-be duties. A short time later we adopted him. When the pyothorax hit, Loki was only three years old and in otherwise good health. We had fallen for this quirky, dropout dog and gave the vets the go-ahead to see if they could save him from his overactive immune system.

Loki's treatment consisted of an oral and intravenous cocktail of antibiotics, anti-inflammatories, and fluids. But his lungs hadn't been properly inflating and on the third day in the hospital, he came down with a second infection—pneumonia.

When we saw Loki at the veterinary hospital, he could barely move. A flexible tube ran from each nostril to an oxygen tank, but he still struggled just to breathe. Everything that epitomized this little Lab had utterly vanished. Lying half-conscious on a blanket, his eyes flickered when we called his name and then fell shut. All he could muster was a limp thump of his thick black tail. Sturdy drainage tubes protruded from between Loki's ribs. The widest Ace bandage we had ever seen wound snugly around his torso to hold the tubes in place. Every few hours veterinary technicians would use these tubes to suction out the fluid and pus collecting in his thoracic cavity.

We waited powerless while Loki teetered on the edge of death. Thirty-six hours after the pneumonia hit, an additional round of antibiotics began to work. Loki rallied and we were ecstatic. The vets were amazed.

FEVERISH CHANGE

Prior to antibiotics, infections and infectious diseases were the biggest threats to human health. For centuries, tuberculosis, smallpox, and typhoid reigned as major killers of people around the world. Outbreaks of yellow fever plagued New York and Philadelphia in the late 1700s and early 1800s. About 5 to 10 percent of those who became infected died. But surviving meant suffering—fever, aches and pains, and telltale yellowing of the skin and whites of the eyes. By 1822, mosquito-control efforts largely quelled yellow fever in northern cities, but southern cities remained plagued into the late 1800s.

As yellow fever began to wane in the North, a new disease arrived on ships from Asia. In 1832, cholera broke out in New York and surged through large U.S. cities for decades. Each epidemic brought cities to their knees. Those with the means to flee to the countryside did, but the poor, mostly recent immigrants, remained behind and suffered.

The toll of infectious diseases on Americans gradually lessened as drinking water and sewage systems, garbage collection, and other sanitation measures came into widespread use. By the 1940s, the development of antibiotics and vaccines had begun in earnest. Together with continuing public health measures, these new treatments brought one notorious pathogen after another under control.

But modern scourges, the so-called chronic diseases such as arthritis and juvenile (type 1) and adult-onset (type 2) diabetes, proliferated in the decades after the Second World War. Chronic diseases can't be passed from one person to another and they can strike anytime from childhood to adulthood. And once you have one, it's difficult to shake. According to the Centers for Disease Control and Prevention, chronic diseases accounted for seven of the top ten causes of death among adults in America in 2010. And, the agency reports that in 2012 about half of all adults had at least one type of chronic health condition.

Constant, low-grade inflammation is a hallmark of chronic diseases. But unlike Loki, whose symptoms of inflammation due to his fast-moving bacterial infection were overt, a person is usually unaware of low-grade inflammation. Imperceptible inflammation, however, is not innocuous—over time it can cause life-threatening problems.

Many chronic diseases also have an autoimmune component. Autoimmunity arises when a person's immune system mistakes one's own tissues as something that shouldn't be in the body. The normally helpful inflammatory process runs amok and destroys or damages healthy tissue. Autoimmune diseases can occur nearly anywhere in the body: multiple sclerosis harms the nervous system; some types of asthma damage lung tissue; and Crohn's disease strikes in the mid to lower reaches of the intestines.

Although there are more than eighty different types of autoimmune conditions, not all have been well studied, or consistently tracked, which leads to variability in prevalence estimates. The National Institutes of Health estimate that about 8 percent of the population lives with an autoimmune disease, while the American Autoimmune Related Diseases Association estimates that more than twice that number do.

In part, the rising incidence of chronic and autoimmune diseases is due to fewer deaths from infectious disease in early childhood and better care should one strike. And today, certain diseases, like asthma, have treatments that did not exist in previous decades. In addition, the increased incidence of many autoimmune diseases is due to both better detection and diagnosis as well as a reassignment of some chronic diseases to autoimmune diseases. These factors, however, still don't account for the increasing number of people who have developed chronic and autoimmune diseases in recent decades, especially children and young adults.[1]

The occurrence of autoimmune diseases rose dramatically in developed countries in the last several decades of the twentieth century and shows no signs of abating. In the developed world, type 1 diabetes has more than doubled in the last thirty years and is striking at an ever-younger age. Modern health ailments linked to chronic inflammation are now as much a threat as our oldest foes, the pathogens.

THE DUALITY OF IMMUNITY

Searching for explanations and answers to the puzzling rise in chronic and autoimmune diseases necessarily leads to the immune system and its curious duality—that it can both save us and harm us. But the immune system is also hard to envision. There is nothing big and flashy about it like the brain of our nervous system or the heart of our circulatory system. Much of our immune system spreads out all over the body as a diffuse network of lymph vessels and lymph nodes. Immune cells travel around the body in this network as though we were one vast sea.

In addition to the GALT, there are other major players in our immune system. How many of us even know of the places from which our immune cells hail—the thymus, and strangely, the marrow inside of our bones? The walnut-sized thymus sits just below the breastbone and gives rise to one type of immune cell, while our bone marrow is the womb of another kind. And what about our fist-sized spleen? It sits tidily tucked beneath the rib cage in the upper left abdomen, filtering our blood to screen out foreign molecules.

Most of us think of the immune system solely as our body's defense system. This standard view focuses on what we'll call barbarians-at-the-gates, shorthand for how the various immune cells battle pathogenic microbes. The immune system has superb detection and recognition skills. It knows that barbarians are not part of you and generally stomps them out.

But our well-being requires more than just beating back the barbarians. Much more. There is, after all, a vast world inside the gates. It consists of the internal environment of the body and its major parts, the heart, lungs, liver, gut, kidneys, brain, and so on. And it is here, behind the gates, that the immune system plays its other critical role—regulating the overall level of inflammation in the body so that all systems can operate trouble-free and seamlessly with one another.

Pinning down exactly how the immune system optimizes and continuously manages its double-edged sword of inflammatory power has proven complicated. Immune cells as a group are incredibly diverse in what they do and what they look like. And confusingly, some of the same immune cells that fight barbarians also keep order inside the gate. Biologists have some ideas about why our particularly complex immune system arose.

One holds that an immune system endowed with serious firepower is necessary for us and other long-lived mammals because over a lifetime we will encounter more pathogens and parasites than, say, a grasshopper or a frog—sometimes more than once. Thus, the argument goes, we require a more robust and sophisticated defense system. But this idea fails to explain why some invertebrates with relatively simple immune systems can live over a hundred years, like quahog clams that filter bacteria-rich sediments.

Another emerging view of the immune system is rooted in the observation that the gut microbiota of vertebrates consists of relatively stable, highly diverse communities with a large number of species that live nowhere else. These traits stands in stark contrast to the highly transient communities of microbes that live within invertebrates and constantly change with external environmental conditions. Our complex immune system allows us and all other vertebrates to identify all the microbes we come in contact with—barbarians *and* our lifelong resident microbiota.

If we accept that our bodies are not entirely our own, we can reconcile the dual nature of our immune system. Consider the many microbial landscapes composing our bodies—from the river valley of our gut, to forests of hair, dry toenail deserts, and the skies of our eyes. These places have a multitude of interacting inhabitants and are as dynamic as any other ecosystem on Earth. They are also subject to the routine natural processes that occur in visible ecosystems—cycles of resource abundance and scarcity, catastrophes, predator-prey relationships, temperature and moisture gradients, and so on.

It is in our best interest that the inhabitants of our bodily ecosystems do not trigger our immune system to adopt a scorched-earth policy. This would be ruinous for both us and our microbiome. And so, consider that the underappreciated, full-time job of our immune system is to maintain the health of the body's myriad ecosystems and their inhabitants. There will of course be the need for occasional bursts of inflammation to destroy barbarians who show up at the gates, but overall the paramount purpose of the immune system is to make sure our bodily ecosystems work as they should—in our interest.

Such a view squares pretty well with new microbiome discoveries. Evidence is accumulating that the mammalian immune system has evolved to keep tabs on and maintain peace with the long-term microbial residents of our bodies. And as in any symbiotic relationship, when they thrive, we too thrive. Although mechanisms and details are not entirely known, disrupting the microbiome appears to lie at the heart of our susceptibility to many chronic and autoimmune diseases.

TOO MUCH OF A GOOD THING

To fully grasp the significance of what scientists are discovering—daily, it seems—about the human microbiome, it is important to have some grounding in how the immune system works. Though diffusely spread throughout our bodies, the immune system is akin to a switchboard-like interface across which all microbes we encounter communicate with us, and we with them.

Swollen lymph nodes, achiness, or a fever are among the telltale signs that your immune system is battling a pathogen. And as we've mentioned, inflammation is normally a sign of health. Say you cut your finger while chopping carrots for dinner. No matter how clean, the knife, the carrots, and the cutting board all harbor bacteria. Once the blade breaks the skin, bacteria rush in, and your immune system just as quickly goes to work. Nearby blood vessels become purposefully leaky, allowing immune cells to charge out of the bloodstream and join other immune cells resident in the skin. The immune cells at the wound site secrete compounds called cytokines to communicate with one another and with other immune cells farther afield.[2] Some of the immune cells called to the wound site kill the bacteria that entered your body, and others begin the process of healing the wound.

Inflammation manifests as redness and tenderness around a wound because of blood and immune cells drawn to the injury. Later on, certain cytokines prompt the skin cells and blood vessels around the cut to regrow and patch up the area through which the knife cut. The immune cells that create inflammation are like an all-in-one wrecking, remodeling, and cleanup crew, intent on getting into and out of a building as fast as possible. So acute inflammation, at its essence, is an important part of the healing process.

But low-level chronic inflammation is a different story. In this scenario the firepower of the immune system takes aim at perfectly normal, healthy tissue. Immune cells mistakenly think a wound needs healing and release a nonstop flood of cytokines, generating even more immune system activity. Immune cells at the site of inflammation keep lobbing destructive substances and don't know when to quit. The commotion feeds on itself, spawning a sort of biological chaos. Over time, conceivably years, tissues subjected to chronic inflammation become a festering mess.

The cytokines and bevy of immune cells that flock to the site of a wound or infection make the local environment a *very* busy place at the cellular level. Turn this process on day in, day out and it becomes a liability. With no wound to repair, and therefore no end point in the inflammatory process, immune cells and cytokines stick around and fuel an unusually high rate of cell division. Every time a cell divides, its DNA is copied. But sooner or later copying mistakes occur. Some mistakes do not matter, or can be fixed, while others have profound consequences, like propelling a few cells down a path of uncontrolled growth—toward cancer.

TWO ARMS

Immunologists generally refer to the immune system as having two parts—the innate arm and the adaptive arm. Each works in distinctly different ways. "White blood cells," a phrase you are likely familiar with, is the general name for all immune cells regardless of whether they hail from the innate arm or the adaptive arm. But particular types of immune cells are associated with each arm, and take on the same general name, innate or adaptive. Both types of immune cells circulate through the body via the interconnected system of lymph and blood vessels.

The first task of your immune cells is to scan and identify the innumerable and diverse kinds of molecules they encounter. They have a singular purpose—to establish if the molecules are "you" or "not-you." Are these molecules from a barbarian? Or, are they citizens of your microbial republic?

Immune cells also keep tabs on all the cells of the body, because sometimes, as with cancer, cells go bad. When everything is running smoothly, the immune system recognizes and distinguishes the different molecular signatures associated with pathogens and nonpathogens as well as those from our own cells.

There are some basic distinctions between innate and adaptive immune cells. The cells that compose the innate arm of the immune system can recognize most common pathogens right off the bat. Innate immune cells are our first line of defense. "Innate" refers to the fact that these cells are part of every person at birth.

Thus far, immunologists have found more than a dozen special receptors on the outer surface of innate immune cells that can detect different kinds of microbes. Each type of receptor is much like a built-in bar code reader but with a key difference. The receptors of innate immune cells scan for broad patterns—the molecular signatures associated with different pathogenic and nonpathogenic microbes.

What this business of receptors boils down to is that innate immune cells possess first-rate detection skills similar to those of search and rescue dogs laser-focused on a target scent. And this brings us back to the scanning that certain innate cells perform. When they encounter a molecular signature they recognize, they collect a sample and show it to immune cells from the adaptive arm. The molecular samples that dendritic cells collect are so important that they have a special name—antigen. Without antigen the adaptive arm of immunity would be clueless about what to do.

Macrophages and dendritic cells are types of innate immune cells called phagocytes ("cell eaters"). While they can both detect the molecular signature of common pathogens, they do so for different reasons. Like a true predator, macrophages hunt so they can eat, and in the process they kill the hapless microbes they engulf. Dendritic cells are quite different. Their purpose isn't to kill a pathogen, but rather to obtain antigen from the intruder. Both dendritic cells and macrophages are capable of recognizing certain molecular signatures found in microbial pathogens but not in people.

In contrast to innate immune cells, adaptive immune cells don't have automatic recognition and killing powers. They need information from innate immune cells before they will act. There are two main types of adaptive immune cells—T cells and B cells. One of the most remarkable aspects of T cells and B cells is that once an innate cell presents them with antigen, they can remember it for the rest of your life. If you pick up the same bug a second time, the T cell or B cell is onto the pathogen instantly, which greatly reduces symptoms and can even mean the difference between life and death. This extraordinary characteristic of adaptive immune cells is why vaccines work.

As remarkable as your immune system is, it can be tricked. Cancer cells have maddeningly clever ways of doing so. Sometimes cancer cells display telltale signs that they are abnormal and immune cells respond with all their firepower. But cancer cells can evade immune cells at every turn, sometimes secreting compounds that bollix cytokine messaging, thereby derailing the immune response. In other cases immune cells may read cancer cells as partly you and partly not-you, so some immune cells attack the cancer cells and other immune cells call off the attack.

The HIV virus presents an even more pernicious challenge for our immune system. It infects a linchpin type of immune cell on which nearly every type of response in our immune repertoire depends. In the end, this spells disaster, making people with HIV increasingly vulnerable to common ailments.

INTREPID EXPLORERS

In the early 1970s, Ralph Steinman, a young immunologist at Rockefeller University, began looking into dendritic cells. Initially, though, he did not know what they were. In the course of experiments, however, he came across strange cells with arm-like features of various lengths. No two cells appeared the same. Some looked like stars and others like stretched-out blobs. More surprising, these cells could extend and retract their arm-like features and change their shape like an amoeba. Steinman named his marvelous new find a dendritic cell and studied it for decades.

He eventually learned that these odd cells are intrepid explorers, sailing the antigenic seas of the body collecting samples that play a central role in initiating an immune response. In fact, immunologists now know that without dendritic cells there would be a disastrous breakdown in communication between the two arms of the immune system.

Later in life, Steinman developed pancreatic cancer and became his own test subject. He worked with other immunologists he knew and developed experimental treatments for his cancer. His star ally in this endeavor? The dendritic cells he had discovered. Steinman lived for four and a half years after his diagnosis, far longer than the weeks or months typical for most people with the same cancer.

On October 3, 2011, three days after he died, the Nobel Committee called to tell Steinman he would be sharing a Nobel Prize for his work on dendritic cells (the other half went to a pair of scientists working on the activation of innate immunity). In more than a century of handing out Nobels, such a thing had never happened. Although their rules bar posthumous awards, the committee let Steinman's award stand. Until they called, they thought he was alive.

Almost a century before Steinman studied dendritic cells, and later their role in fighting cancer, Dr. William Coley stumbled onto a way to cure cancer using bacteria. Coley was an assistant surgeon and instructor of surgery at New York Cancer Hospital (today's Memorial Sloan Kettering). One of his colleagues recounted the stunning turnaround of a thirty-one-year-old patient with a cancerous neck tumor who had been operated on five times in three years. After each operation, the tumor grew back. Unable to remove the entire tumor in the fifth operation, Coley's colleague gave up, certain that the man was doomed. But, shortly after the final surgery, the patient contracted erysipelas, a skin infection unrelated to cancer. The erysipelas raged, diminished, and then recurred a short time later in milder form. Astonishingly, the man's tumor began to shrink and then disappeared completely. Seven years later, Coley and his colleague reexamined the man: the neck tumor had not grown back!

Coley was fascinated by the connection between the skin infection and the disappearing tumor. He wondered what effect inducing an erysipelas infection might have on other patients with similar types of cancer. On May 2, 1891, he got the chance to find out. A patient was referred to him with neck and tonsil tumors too large to operate on. Coley injected the man with a bacterium he thought would cause erysipelas. It didn't work. Coley suspected he had the wrong microbe.

Determined to find the right one, he contacted German microbiologist Robert Koch, who arranged for the transfer of a pure culture of a true erysipelas-causing bacterium across the Atlantic Ocean (no small feat in those days). Escorted by a professional acquaintance of Coley's, the pathogenic *Streptococcus* strain arrived safely in New York City. Coley promptly inoculated his patient. This time erysipelas did set in, and a couple days later the neck tumor started to break down. In two weeks, it was gone. And after two years the neck tumor had still not grown back (although the tonsil tumor stayed about the same size).

Coley continued using the *Streptococcus* strain from Koch with mixed results. He did not fully understand how or why he achieved the results he did, but it seemed that the bacterium that triggered erysipelas also activated the part of the immune system that could detect and kill tumor cells. Although the mechanisms eluded him, Coley was developing an early version of immunotherapy to harness the know-how and power of the immune system to fight disease.

THE LANGUAGE OF ANTIGEN

Steinman's discoveries led to a deeper understanding of Coley's results.

It turns out that the strange shape of a dendritic cell is as perfect for what it does as the wings of a bird are for flying. Recall that many of the cells and much of the tissue composing the immune system wrap around the digestive tract. Of all the different kinds of immune cells, only a dendritic cell can go from the outside of the colon to the inside. It nimbly slips one of its starfish-like arms in between two colon cells to start off its fishing expedition. When the arm reaches the layers of mucus, it dives through like a submarine's periscope and surfaces on the inside of the colon. The arm of the dendritic cell collects antigen from the colon lumen or the mucus and leaves the same way it entered.

Dendritic cells also come equipped with a special molecule that functions like a flagpole rising up from the cell surface. They run antigen up their flagpole for display purposes. When researchers realized what dendritic cells could do, this unlocked the secret of how the two arms of the immune system communicated with each other and with microbes. Dendritic cells and T cells rely on a common language—antigen.

The colon is but one of many places that dendritic cells monitor. They congregate in the places that the potential exists for microbes from the external environment to enter our bodies—like the skin, lungs, and if you have one, the vagina. In this way, dendritic cells are not just explorers, they are expert at cruising our bodily seas to load their hold with antigen to show T cells.

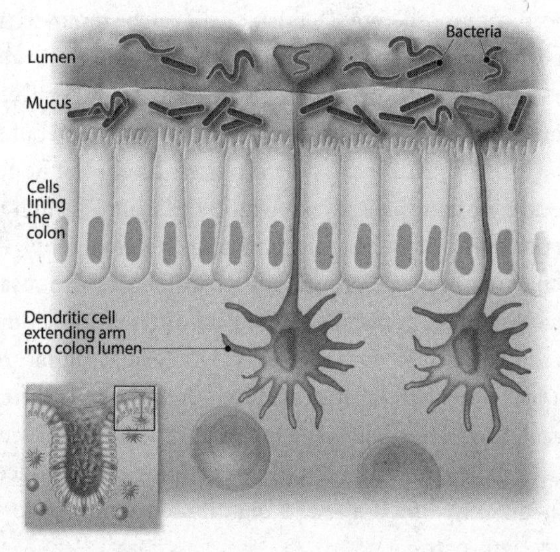

A Living Periscope. Dendritic cells collect antigen from the colon to show it to other immune cells.

Although the birthplace of all T cells is the thymus, that is not where dendritic cells and T cells encounter one another. Nor is a T cell considered active upon departing from the thymus. The first phase of life for a T cell is to roam the vast seas of the body. T cells have receptors that are specific for different kinds of antigen and they repeatedly stop over in the harbors of the spleen and lymph nodes to meet up with dendritic cells and inspect their cargo of antigen.

A T cell will activate only after a particular set of events occurs. First, the T cell and dendritic cell must actually find one another in the labyrinth of a lymph node (or the spleen). Second, the dendritic cell must be displaying the antigen for which the T cell receptor is made. When both things happen the T cell activates. And once activated, whether during childhood or adulthood, a T cell generally sticks around for the rest of your life doing the same thing over and over again.

There are at least a half dozen kinds of T cells, and each type serves our immune system in its own way. Killer T cells, for example, search for the mother lode of antigen that activated them—primarily cancerous or virus-infected cells—and then kill them.

The ability of dendritic cells to activate killer T cells is the basis for one form of cancer immunotherapy. Dendritic cells can be removed from a person and loaded up with antigen from their tumor. Once introduced back into the person's body, the dendritic cells will activate killer T cells, which go on to search out the antigen source (the tumor cells) and kill them.

In contrast to T cells, B cells start out their life in the bone marrow. Like T cells, they roam the body, constantly passing through the spleen and lymph nodes in search of antigen that will activate them. Activation of a B cell culminates in the production of antibodies—thousands a day when needed. Antibodies circulate freely throughout the body in blood and lymph fluid and "tag" microbes whose molecular signature matches the antigen that activated the original B cell. A microbe with an antibody tag on its back quickly becomes the target of other immune cells, which will then kill it. The strength of B cells is that they can produce antibodies relatively quickly, which makes a big difference in countering a fast-moving infectious pathogen.

BALANCING ACT

While B cells are an important part of the immune system, researchers have uncovered some particularly surprising and interesting ways that T cells interact with our microbiome.

Up to now we have mostly paid attention to interactions between the immune system and pathogens. But it is important to remember that not all microbes are pathogens. In fact, bacteria in the human gut microbiome are mostly helpful, not harmful! This reality goes a long way toward explaining why microbiome researchers are discovering that interactions between immune cells and nonpathogenic microbes are critical to normal immune responses.

In particular, two types of T cells—regulatory T cells (Tregs) and Th17 cells (so named for interleukin 17, a cytokine it secretes)—seem to take cues from nonpathogenic bacteria. These two types of T cells protect you in a different way than killer T cells. They don't attack pathogens or tumors. Rather, they carry out the second major role of the immune system, the day-in, day-out routine of meting out inflammation. Tregs quell inflammation and Th17 cells fire up inflammation.

In people who have a normal immune system, these two types of T cells are often depicted like two perfectly balanced weights perched on either side of a seesaw, playing a unified master-regulator role. Together they dial the level of immune firepower up or down to achieve the optimal level of inflammation needed to battle a pathogen, tolerate a helpful microbe, or heal an injury.

Under certain circumstances, proinflammatory Th17 cells weigh down their side of the seesaw and disrupt the balance. An imbalance between Th17 cells and Tregs contributes to autoimmune diseases and inflammatory ailments like ulcerative colitis as well as inflammation-driven cancers.

The way all the different T cells make receptors hinges on one of the most fascinating biological processes. While it's considerably more complicated than this, let's say the end of every T cell receptor is like a unique hand. Each such hand has a different number of fingers and a different size and shape. Imagine this hand waving about, outstretched in search of its mirror hand, the antigen displayed by a dendritic cell. The T cell receptor must find its precise match, though. And there are a lot of potential matching hands out there—millions of them. So when a dendritic cell shows up with antigen, a T cell receptor reaches out, so to speak, and does an exacting check of the fit between its receptor and the antigen. When the fit is as perfect as two puzzle pieces, the T cell snaps into action.

But one killer T cell is no match for a throng of cancer cells, let alone cells brimming with pathogenic viruses. Nor can one Treg or one Th17 cell orchestrate holding back a tide of chronic inflammation. So, an acti-

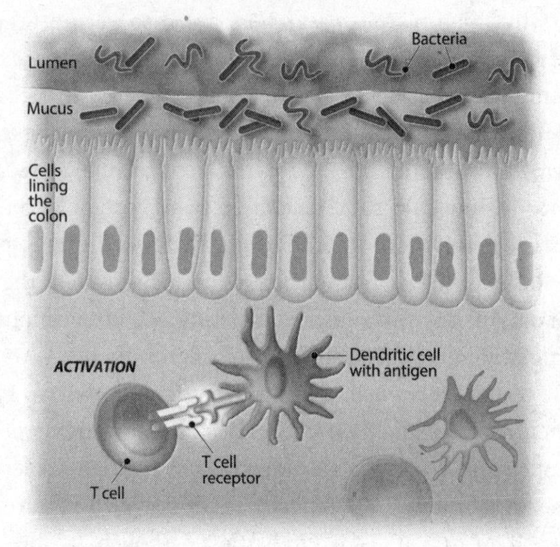

Activating T Cells. A T cell checks the "fit" of antigen on a dendritic cell. Only the right kind of antigen will activate a T cell.

vated T cell's first order of business is to clone itself over and over again. The cloned T cells will have receptors identical to those of their T cell progenitor.

Our bodies generate T cell and B cell receptors through an ingenious process. Scientists initially thought that the genes that coded for T cell and B cell receptors were inherited. But, the incredible diversity (and thus number) of immune-cell receptors would require that nearly all of our genome be dedicated to that purpose. Instead, the gene to make a receptor is assembled on the spot, through the random rearrangement of a far more modest number of DNA segments. In this way billions of different receptors are generated.

While plants manufacture an immense array of phytochemicals and exudates to attract and retain allies, we have the analytic prowess and communication skills of dendritic cells. We can also assemble an astounding diversity of T cell and B cell receptors. This is exactly what we need to productively interact with microbes from across the tree of life.

MICROBIAL ALLIES

Although Americans routinely died from microbially caused ailments just a century ago, infectious-disease scientists estimate that there are actually relatively few human pathogens, around 1,400. In contrast, estimates put the number of nonpathogenic microbes in humanity's collective microbiome at about a million. That's nearly 700 types of nonpathogens for every pathogen. Moreover, microbiologists think they may have identified only about 1 percent of all the nonpathogenic microbes that are out there. And consider that whether pathogenic or helpful, many microbes can tweak their genes on the turn of a dime.

Adding to the surveillance our immune system conducts is another twist that has to do with microbes themselves. A fair number of bacteria with whom we coevolved wear more than one hat. These hat-changers can move along the symbiotic continuum from the mutualist end toward a neutral middle position or slide down to the pathogen end. But the majority of the time, our permanent microbial residents help us. At other times they are neutral and infrequently they harm us. Environmental factors, say a different food source, or a new microbe entering or leaving the community, are among the reasons that bacteria change hats. The microbes in our microbiome that behave this way are called commensals.[3]

As you may recall, the majority of the immune system is intertwined with our digestive tract, especially the colon. So why don't the immune cells that hang around the colon attack such a microbe-rich cesspool? After all, this churning mass of partially digested food attracts the attention of snoopy, nosy dendritic cells that periscope into our gut to collect antigen.

One explanation is that immune cells simply ignore commensals. In this can't-see-you hypothesis, the thick layers of mucus that coat the luminal side of colon cells are thought to act as an impenetrable wall separating commensals from both the colon cells themselves and the net of immune cells and tissue surrounding the colon. If microbial antigen never makes it through the mucus to the immune cells clustered on the outer side of the colon cells, then an immune response can't be triggered. This explanation makes a lot of sense, but only if you make two incorrect assumptions—that the immune system only interacts with pathogens and that commensals stay in the lumen.

Let's put some of these startling facts together. We omnivorous, big-brained, long-lived mammals have waded through a world of microbes ever since we came into being. Microbial cells significantly outnumber our own bodily cells. The number of nonpathogens handily swamp their pathogenic brethren in the environment inside and outside of our bodies. Our genome possesses an on-the-spot ability to generate receptors that detect an astronomical amount of biological diversity. Our immune cells endow our immune system with spectacular surveillance powers to sort pathogens from allies. What these facts point to is that the story we've told ourselves about immunity—its purpose, where it comes from, and how it works—is in for a plot twist.

What if our immune cells aren't so much an army as linked-at-the-hip environmental sentinels? And what if their day-to-day job in our life is to maintain a standby level of inflammation that is not only optimal for all of our major systems—but also for the microbiota that help run them. Then, when a once-in-a-while pathogen does show up, our immune sentinels can quickly fire up the appropriate inflammatory response to oust the intruder.

The challenges of staying alive today are clearly different from those we faced when our modern ancestors evolved about 200,000 years ago. But one thing has remained constant—we live in a microbial world filled with duality and complexity, inside and outside of us. A pathogen-centric view of our immune system is as wrongheaded as the pre-Copernican view that the Sun orbits Earth.

Today's microbiome researchers are having a heyday that rivals that of globe-trotting nineteenth-century naturalists like Charles Darwin and his adventurous peers Alexander von Humboldt and Alfred Russel Wallace. These explorers discovered one natural wonder after another—from the heights of the Andes to the jungles of Borneo. Upon their return to Europe, their specimens and ideas forever changed the way we see the world. Now, two hundred years later, scientists are exploring inner landscapes—ecosystems and inhabitants of the human body—once again challenging our view of who we are.

SEEDING COMMENSALS

Like plants, we tap into the nature of our immediate environment to assemble and cultivate our microbiome. But our acquisition plan is a little more complex. In the hours before birth, our mothers ramped up production of a special type of vaginal mucus and grew a unique crop of microbes—just for us. As we slipped out of the womb and began our descent into the world, they latched on. We the hand and they the glove; it was the fit of a lifetime in many ways. And near the very end of our birth journey, a daub of mom's feces added the final touches to our initial microbiome. Now some scientists think the placenta and perhaps even the uterus may help seed us with our initial microbiome, as bacteria from mothers has been found in umbilical-cord blood and amniotic fluid.

The true significance of commensals is brought to light in reviewing the numbers again. Recall that microbial cells greatly outnumber our own, especially in our gut. And while a bacterium weighs just one millionth of a millionth of a gram, all together your microbiome weighs several pounds. A square inch of human skin hosts around half a million microbes—roughly the population of Wyoming. There are more microbes in you than stars in our galaxy, the Milky Way. Far more. We are each our own microbial galaxy. And, in the case of bacteria, the overall mix of microbes that composes a person's microbiome is not only as unique as their fingerprint, but it changes over time. The microbiome of a fifty-year-old doesn't much resemble their two-year-old microbiome.

And remarkably, like the bacteria that inhabit the rhizosphere and communicate the presence of pathogens to a plant, there is evidence of similar activity occurring in the colon. Bacteria that live in the mucus layers can sound the alarm to colon cells via chemical messaging when pathogens from the lumen attempt to colonize the mucus layer.

Some commensals benefit us so much that we suffer when they're absent. While it's long been known that pathogens trigger an immune response, it's now becoming just as clear that commensals interact with the immune system—all the time, not just once in a while. In fact, commensals appear to play as big a role in priming and training immune cells as do pathogens. In some ways, their role is even more important given that microbiome researchers are discovering that commensals play a key role in regulating the overall level of inflammation in the body, which in turn is necessary to keep everything else in our body running smoothly.

THE CURIOUS CASE OF *BACTEROIDES FRAGILIS*

Cal Tech microbiologist Sarkis Mazmanian is among the microbiome researchers building a compelling case that commensals help regulate inflammation. His research group's experiments are unraveling the manner in which the gut dweller *Bacteroides fragilis* interacts with the immune system. Although mice are the experimental subjects, their results point to ways that *B. fragilis* and other mucus-dwelling bacteria found in the human colon may affect our own immune responses. But before we go into Mazmanian's discoveries, it's helpful to review the thinking at the time in which his discoveries occurred.

For a long time it was assumed that dendritic cells could only display antigen made of protein fragments, which led to the idea that T cell receptors—all the millions of them on killer T cells, Tregs, and Th17 cells—could only match protein fragments. After all, protein fragments were all scientists ever found when they analyzed antigen. But *B. fragilis* proved otherwise.

B. fragilis makes a special molecule called polysaccharide A (PSA). Polysaccharides are not proteins; they are carbohydrates. The discovery that dendritic cells can run PSA antigen up their flagpoles meant that dendritic cells could detect and share a broader range of antigen with T cells than previously thought possible.

In fact, the relationship between *B. fragilis* and the immune system hinges on PSA. And this is where Mazmanian's discoveries enter the scene. Mazmanian figured that if dendritic cells bothered to display antigen from *B. fragilis* then there had to be a good reason. He and his colleagues set out to learn more about this common bacterium. Why would a dendritic cell be able to pick up PSA, and why would this molecule be of interest to a T cell?

As you now know, inflammation is a major part of the immune response, and that's where Mazmanian and his crew started to investigate the influence of *B. fragilis* on its host. In classic laboratory fashion, they gathered mice together and designed a series of experiments. The particular mice they used were bred and housed to be "germ-free." This made it easier to study the effects of *B. fragilis.*

First, a pathogenic bacterium known to cause colitis (inflammation of the colon) was introduced into the germ-free mice. Once the colitis set in, *B. fragilis* was introduced and the colitis-stricken mice were soon up on their little feet again. In another group of mice, also afflicted with colitis, a different form of *B. fragilis* was introduced, one that had been genetically modified and could no longer make PSA. The colitis in these mice continued unabated. These experimental results revealed two things—*B. fragilis* could make colitis disappear and the PSA molecule that *B. fragilis* made seemed to play a key role in the cure.

But how exactly did *B. fragilis* cure colitis? Mazmanian and his colleagues kept digging to find the mechanism. It turned out that dendritic cells loaded with PSA antigen activated the anti-inflammatory Tregs! *B. fragilis*, it seems, helps to keep the peace in its mice hosts while avoiding becoming a target of the immune system itself. In doing so, this bacterium keeps a roof over its head. Was this a case where every body wins?

Mazmanian firmly established the anti-inflammatory effects of *B. fragilis* by doing one more thing. He extracted and purified PSA from *B. fragilis* and gave it to mice afflicted with experimental colitis. The effect on the mice was the same as if *B. fragilis* itself had been introduced. Sure enough, their colitis cleared up, and they too were soon back on their little feet.

Mazmanian and his colleagues dug deeper and learned more about the Tregs that *B. fragilis* had induced. These particular Tregs cured the colitis through secreting a cytokine called interleukin 10 that quells inflammation. In mice still afflicted with colitis, a different T cell and cytokine were at work. The mice with colitis had relatively high numbers of Th17 cells, which made inflammation-inducing interleukin 17.

Although *B. fragilis* can cause problems should it colonize places other than the digestive tract, the experiments that Mazmanian and colleagues conducted showed that *B. fragilis* has curative powers, at least in the guts of mice.

Japanese researchers were also curious about the human commensals living in colonic mucus. Could other bacteria besides *B. fragilis* influence the development of Tregs? Through a series of experiments, they tested the ability of different commensals to cure germ-free mice afflicted with colitis. The mechanism, they hypothesized, was through inducing Tregs to develop. They hit on a surprising finding—the biggest booster of Treg development wasn't just a single bacterium. It was a specific mix of seventeen strains from a group of bacteria called Clostridia.

GETTING IT JUST RIGHT

It seems that if ever there was a case to be made for the Goldilocks analogy, it's here—in how the immune system revs up and powers down Treg and Th17 cells to achieve just the *right* amount of inflammation. It's clear why Tregs are a good thing—they quell gut inflammation through secreting an anti-inflammatory cytokine that dampens Th17 cell production. But remember that inflammation is also necessary to heal tissue and oust pathogens. So you'd always want some type of pro-inflammatory T cell teed up to start the healing process or sound the alarm if a pathogen gets into the gut. But what primes Th17 cells to a ready state?

Enter segmented filamentous bacteria, a group of commensal gut inhabitants that have been found attached to the gut wall of many vertebrates and invertebrates.[4] These bacteria prompt the development of Th17 cells, but not a Noah's Flood of them—just enough to perk along well below the boiling point of an all-out inflammatory response. In another experiment researchers from the United States and Japan introduced segmented filamentous bacteria into germ-free mice. The researchers waited two weeks and then introduced a highly virulent pathogenic bacterium called *Citrobacter rodentium*. This bacterium causes serious gut inflammation similar to that caused by pathogenic forms of *Escherichia coli* (*E. coli*) in people. The mice inoculated with the segmented filamentous bacteria experienced markedly better outcomes. They developed far less inflammation-caused damage in their colon and *C. rodentium* did not penetrate into their gut wall.

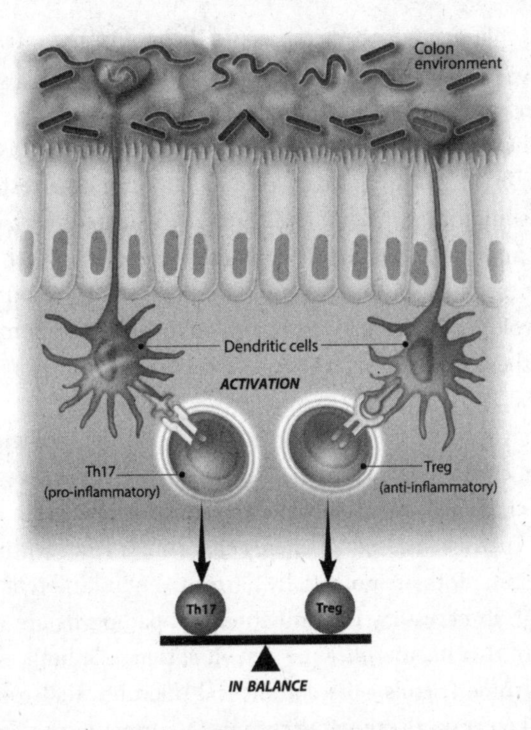

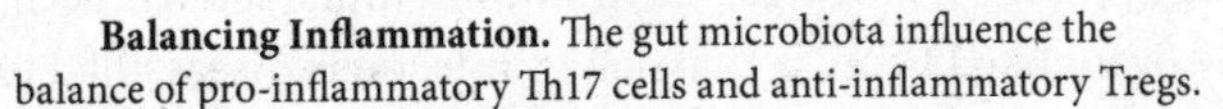

Balancing Inflammation. The gut microbiota influence the balance of pro-inflammatory Th17 cells and anti-inflammatory Tregs.

The mechanisms and molecules underlying the interactions of segmented filamentous bacteria with the immune system remain incompletely understood, as does their ability to avoid triggering the tissue-damaging inflammation that pathogens can cause. Researchers suspect that segmented filamentous bacteria may be prompting a different type of Th17 cell, one that acts more like a subtle tonic than a flamethrower. It's also possible that segmented filamentous bacteria stimulate Tregs to produce cytokines that counterbalance the pro-inflammatory cytokines that Th17 cells secrete, thereby keeping the latter below tissue-damaging levels.

Of course, it's important to have Th17 cells at the ready so they can kick-start the all-important inflammatory process when barbarians reach or breach the gates. Given how quickly pathogenic microbes can reproduce once inside a host (think back to what poor Loki experienced), segmented filamentous bacteria could help ensure a trigger-quick response to oust a pathogen. This can make a big difference in the severity and consequences of an infection.

What the research of Mazmanian and others reveals is that both Tregs and Th17 cells are needed for a healthy immune response. And that, at least in mice, *B. fragilis* and segmented filamentous bacteria help keep anti- and pro-inflammatory processes on an even keel. Commensals like these bacteria finely tune the inflammatory response to help the host avoid debilitating ailments linked to chronic inflammation. In so doing, these kinds of microbes also keep their home hospitable for themselves.

This is an entirely different way of looking at the biological mechanisms that create and support health. Think of it this way: you face two great challenges to your survival—enemies that originate outside your body (pathogens) and an enemy that comes from within (chronic inflammation). Both are potentially harmful, but in different ways.

Although unnecessary inflammation and pathogens are very different kinds of threats, the immune system can handle both, with a little help from some friends—the commensal microbes that roost in your body. Until recently, the threat of chronic inflammation has been almost entirely overlooked because it is not an infectious disease like tuberculosis or cholera. But it's no less a threat, given the connection between chronic inflammation and the rising tide of associated diseases that now occur at epidemic levels in developed countries.

ANCIENT FRIENDS

Once upon an evolutionary time, we routinely encountered an immense diversity of microbial life—in the water we drank and bathed in, on the soil-covered tubers we chewed, and later, in the flesh of animals we consumed. And, even from a source we'd prefer not to think of—our own or another person's feces.

Of course, pathogens could take us down. But once inside the human body, some microbes stuck with the transient lifestyle, passing through our digestive tract, eating along the way, but causing no harm. Some of these commensal microbes even routinely passed from mother to child. Other microbes navigated their way to a truce with the immune system. They became commensals, taking up permanent residence in our gut. Some burrowed deep into pillows of mucus atop the cells lining the colon to stay away from the rushing torrent in the lumen. Or they found their way to some other habitable place within us or on us, and managed to avoid becoming a target of our immune cells.

Consider the appendix. This perplexing backwater at the transition from the small intestine to the large intestine was once thought of as useless at best. Doctors and researchers wondered why this weird appendage-like part of our gut harbored dense biofilms of bacteria and an army of immune cells capable of doing them in.

The answer lies in the surprising function of the appendix. This eddy-like habitat offers commensals safe harbor from the flood-like conditions of the digestive tract. In other ecosystems such a habitat is called a refugium. Some plant and animal species emerge unscathed from refugia after disastrous floods or volcanic explosions, and can recolonize the disturbed area. The community of organisms in a refugium sets the composition of the first arrivals, and when you have a twenty-minute generational turnover time, as you do with many bacteria, having commensals move in right away is of the utmost importance. All of which helps to explain what your appendix is actually good for—a source of ready-to-recolonize commensals after pathogen-purging bouts of diarrhea.

Perhaps the crypts of the colon are another refugia habitat for bacteria. This nutrient-rich niche shielded from the tumult in the lumen is a place from which commensals can quickly recolonize the colon wall when needed.

Wherever they live, the commensals of our microbiome have made themselves invaluable to the immune system and earned a stable home for themselves in return. Most important, over the course of our evolutionary journey, commensals have taught our immune system that they are not pathogens. Instead, commensals offer up the signature currency of the twenty-first century: information. They provide intelligence that helps our immune system avoid triggering unnecessary inflammation.

What this all adds up to is that disrupting our microbiome spells trouble. The sophisticated and precise communication that occurs between commensals, dendritic cells, and T cell receptors is likely to get scrambled. And the last thing we need is for our immune system, endowed as it is with considerable firepower, to be at odds with the body it's supposed to protect.

Botanists have long understood that not all microbes are pathogenic and that the microbial company a plant keeps shapes its fate. Experiment after experiment has shown that sterilizing soils to make them germ-free leads to sick and ailing plants. The same, it seems, is true in the animal world. In three separate studies, germ-free mice exposed to the pathogens that cause dysentery, anthrax, and leishmaniasis (a disease caused by a parasitic protist that results in skin ulcers or dangerous organ swelling) fared far worse than mice with normal microbiota. In addition, germ-free animals also have significantly lower levels of circulating antibodies and produce far fewer T cells in their spleen, lymph nodes, and thymus.

We have never had sterile bodies free of microbial life. And if we were to achieve such a state we would be profoundly unhealthy. The community of microbes that lives within us does myriad things, from helping us oust foes to providing us with their metabolic byproducts, which in turn, help keep us healthy. For example, we are at the mercy of our gut microbiota to make vitamins essential to our health, like vitamin B12, which we need for proper nervous system functioning, and vitamin K, which is involved in blood coagulation and healthy bones. And these are but two of the many molecules and compounds we need in order to live. Microbes produce up to a third of the metabolites found in our blood.

We used to think that the immune system evolved to kill microbes. Now it is looking like microbes help run the immune system. While the details and mechanisms of how beneficial microbes influence our health are only now coming into focus, it is clear that the effects of disrupting one's microbiome can range from annoying to devastating.

What this all adds up to is a need to reimagine ourselves as stewards of our microbiome, keeping our tiny allies well fed, housed, and safe. For when they are well, so too are we. They are not just a hidden part of nature; they are another arm of our immune system—like the third leg on the proverbial stool. And yet, the medical world still largely maintains an antagonistic stance against all microbes, guided by the manifesto of nineteenth-century microbiology—germ theory.

9

INVISIBLE ENEMIES

Those of us born at the tail end of the postwar baby boom and in the years that followed can count ourselves lucky. We received routine vaccinations for diseases that plagued earlier generations. Polio and measles were two diseases common when our parents were children. Go back another generation or two and our ancestors faced diseases that few today could even name.

Although it might seem that epidemics have always plagued our species, that's not actually the case. Long, long ago before the rise of agrarian societies, we lived in small, roving bands of up to about forty or fifty members. People ate whatever they could kill or gather. This kind of lifestyle—constantly on the move and isolated for long periods from other groups of people—prevented infectious diseases from becoming epidemics. Seemingly minor injuries like a wrist or ankle sprain were more of a problem. If you couldn't gather food or dig it up, you might not eat. If you couldn't run, you risked becoming a meal yourself.

It's widely thought that infectious diseases first gained a solid toehold between 5,000 and 10,000 years ago with the rise of early agricultural communities. At the time, farming existed in the fertile Tigris and Euphrates river valleys, as well as in parts of China. Settling down to grow crops and raise animals provided people with a more predictable, and at times more abundant, food supply than did hunting and gathering.

It also resulted in women having more children in a shorter span of time. Imagine the stress of looking after a toddler while constantly on the move, searching for food and avoiding predators. A more sedentary lifestyle allowed women to go from bearing a child every three or four years to giving birth every year or two. While famine could, and did, knock back civilizations, the human population continued to grow. And as our numbers increased, we settled into an agrarian lifestyle, with population densities reaching anywhere from ten to a hundred times higher than that of hunter-gatherer groups.

Squalor was as rampant in ancient times as electronic devices are today. Filthy drinking water and a domestic environment laden with everything from dead animal carcasses to human waste created fertile ground for pathogens. Civilization was off and running with pathogens racing to keep pace.

Perhaps unsurprisingly, many of the deadliest human pathogens got their start in animals we domesticated or that lived off us or our waste, such as rats, fleas, and mosquitoes. Through DNA analysis, scientists know that many microbes—including those that cause smallpox, whooping cough, and scarlet fever—made the short hop from these various animal and insect species to us. But where did the pathogens originate in the first place?

From nature, of course. Every tree we cut down and every field we plowed dislodged microbes and their hosts. You can probably hazard a guess about how the average opportunistic, fast-reproducing, gene-swapping bacterium responded to a new situation in its environment. Given access to new hosts, more food, or better protection from their enemies, they jumped at the chance, either perishing or prospering. Concentrations of people offered bacteria and other would-be pathogens the perfect setup—dense intermingling populations. Much the same story applies to early plant pathogens. Cereal crops like wheat grew as monocultures and, if there were slim pickings for pathogens in the surrounding native vegetation, pathogens made the easy jump to crops.

Time and again, diseases caused by bacteria, viruses, and protists swept through societies, delivering fever, death, or food shortages that profoundly influenced history. When people speak of "the plague," they are generally referring to the bubonic plague—one of the worst scourges of humanity. This infamous disease, caused by the bacterium *Yersinia pestis* that lives in fleas, shaped much of history. The Plague of Athens devastated the ancient city-state in 430 B.C., influencing the course of the Peloponnesian War and helping to usher in the end of Classical Greece and set the stage for the rise of Rome. Almost a thousand years later the

Plague of Justinian (541–542 A.D.) struck the Byzantine capital of Constantinople, hastening the end of the Roman Empire. Worst of all was the "Black Death," which decimated Europe in the fourteenth century. Over the course of eight years from 1346 through 1353, the plague killed a third to a half of all Europeans, the aftermath of which resulted in enormous religious, economic, and social change. Centuries later, smallpox and other European diseases devastated the indigenous peoples of the Americas, allowing the conquest of the New World. During the American Civil War and the First World War, battlefield infections are said to have killed more soldiers than direct enemy fire.

The origins of infectious diseases mystified early societies. Since the days of Hippocrates, disease was attributed to miasma, another word for bad air. Such air was thought to emanate from unsavory places, like stagnant water and decaying bodies. This notion held sway for a thousand years, until European natural philosophers started to challenge classical ideas and investigate the workings of the natural world for themselves.

Almost a century before Antony van Leeuwenhoek discovered animalcules so small that a hundred of them could lie end to end across a grain of sand, an Italian doctor proposed that tiny particles caused disease. In 1546, Girolamo Fracastoro wrote that invisible contagion spread sickness from one person to another, either through direct contact or through the air. He did not assert that they were actually living entities, but he did presume that their presence on soiled clothes and linen transmitted sickness and death. Fracastoro didn't realize it at the time, of course, but his ideas laid an early stone on the path to germ theory—the idea that microbes cause and spread diseases—launching a journey that would take centuries to unfold.

Along the way, people learned that improved hygiene and sanitation proved central to taming disease. Basic public health measures like clean drinking water and municipal sewage systems began appearing in the late nineteenth century. Such efforts controlled and eradicated more infectious diseases than any other method since. But even so, some diseases remained major killers. Vaccines provided another powerful strategy. They didn't kill microbes, they boosted immunity. One vaccine in particular stands out as a cornerstone of medical innovation.

POLIO

At the start of the twentieth century, polio epidemics swept across the United States. The polio virus enters a person usually via food or water contaminated with feces. If not ousted by the immune system, the virus moves from the gastrointestinal tract to the nervous system, where it can cause permanent paralysis of one or more limbs, and sometimes death. A 1916 outbreak in the United States paralyzed 27,000 people and killed another 6,000. Such outbreaks continued for decades, occurring nearly every summer.

By the early 1950s, Americans cited nuclear annihilation and polio among their top fears. While the Russians rushed to develop a hydrogen bomb, American scientists raced to produce a polio vaccine. It couldn't come soon enough. By 1952, the number of U.S. cases of polio peaked at 58,000. The disease threatened every American family regardless of income or class.

Although viruses are not considered a life form, vaccine makers use the terms "live" and "dead." A live virus that has been weakened is generally considered more virulent than the same strain of dead virus. The immune system is touchy and, as with inflammation, it's a balancing act for vaccine makers to find and culture strains or isolate antigen that will trigger a sufficient and effective immune response. Among the challenges is the fact that antigen from pathogens is often linked to the very traits that make a pathogen virulent. In the case of a polio vaccine, the decision to use a live virus or a dead one had profound implications for people's lives and careers.

Dr. Jonas Salk chose to make the vaccine from a highly virulent but recently "killed" virus. In early 1954, his team at the University of Pittsburgh School of Medicine crossed the finish line with a vaccine that worked in the lab. That April almost 2 million children took part in a nationwide trial. By March of 1955, the results were in. President Eisenhower and other dignitaries showered Salk with praise, trumpeting the coming defeat of polio. Dr. Salk refused to patent the vaccine, declaring that it belonged to "the people," an attitude that endeared him to his fellow Americans. Colleagues bitterly resented the attention Salk received and complained that he had not fairly credited all who worked on the vaccine.

With summer around the corner, production of the vaccine began in earnest. By mid-April, pharmaceutical companies were distributing hundreds of thousands of doses. The country breathed a collective sigh of relief as a mass vaccination of children got under way. But a few weeks later, it became clear that something had gone horribly wrong. A girl in Idaho who had received the vaccine came down with polio and died. About a dozen similar cases were reported among children who lived up and down the West Coast. Jonas Salk went from hero to villain in a matter of days.

Chief among his critics was another researcher, Albert Sabin. The two had met when Salk first arrived in Pittsburgh. The inner circle of polio researchers was small, and Sabin, who considered the junior Salk an inexperienced newcomer, belittled his work. Interestingly, the two had remarkably similar backgrounds. Both came from Jewish families who had fled their home countries to escape persecution. Sabin's family had come from Poland and Salk's from Russia. Despite their common ground, there was no camaraderie. The two had starkly different ideas for developing a vaccine for polio. Sabin insisted that a different strain with lower virulence that had been weakened but not killed would yield the safest and most effective vaccine. His concerns about Salk's vaccine made from a highly virulent but "killed" strain seemed well justified.

The deaths associated with the rollout of the Salk vaccine understandably alarmed parents, and many refused to vaccinate their children. But, major polio outbreaks again swept through Chicago and Boston with the onset of summer. Because of the scare, few children had been vaccinated. Salk was said to be on the verge of suicide.

As it turned out, one of six commercial labs bungled manufacturing the Salk vaccine. Exactly as Sabin had feared, the problem arose from using a highly virulent strain. During the production process some of the culture vessels sat too long and sediment formed in the bottom of these vessels. The chemical used to kill the virus didn't fully reach all the virus particles at the bottom of the sediments, and hence some live virus made its way into the vaccine. Once discovered, the problem was easily fixed. Vaccination resumed using Salk's vaccine and no more children died. But the damage was done, the seeds of doubt sown.

In the meantime, Sabin persevered, developing a vaccine made from a weakened live virus far less virulent than the strain Salk had used. Many believed that Sabin's vaccine would be safer than and just as effective as Salk's, but it had never been tested. The Russians obliged, desperate to control polio after a devastating outbreak in the late 1950s. At the height of the Cold War, Russian scientists came to the United States to evaluate both vaccines. They chose Sabin's and agreed to organize a trial of 10 million children. The trial results fulfilled Sabin's dream. His vaccine proved safe and the Russian regime moved swiftly to vaccinate another 70 million children.

Back in the United States, the battle continued over which vaccine to use. In 1961, less than a decade after the Salk vaccine came into use, cases of polio dropped to below a thousand. Clearly it worked. But the successful, large-scale effort in Russia caused Americans to question why they weren't receiving Sabin's vaccine. If it was safe and good enough for the Russians, then wasn't it good enough for Americans too? Such discontent played into Sabin's hands. He lobbied government officials, as did drug companies, who stood poised to make tidy profits from the patented vaccine. Salk's protests made no difference. In the fall of 1961, Sabin's vaccine was the one recommended to doctors. Cases of polio continued to drop, and in 1979 it was declared eradicated in America. Globally, however, polio still wavers on the edge of eradication. Just when it appears down for the count, it rises again on tides of war and social instability.

THE POX

The "speckled monster," smallpox, had ravaged civilizations for far longer than polio. While many Americans alive today can still remember the polio outbreaks of the 1940s and 1950s, smallpox had already been eradicated in the United States. Outbreaks continued in developing nations but, by 1980, smallpox became the first disease to be eradicated from the planet.

The virus *Variola major* causes the most fatal form of smallpox. It is among the pathogens that began plaguing humanity early on, shortly after the dawn of agriculture. Smallpox epidemics waxed and waned through the centuries, and the populations with the longest exposure to the virus slowly began to build up immunity.

People of all ages had about a one-in-three chance of contracting smallpox during an outbreak and, once infected, about a one-in-five chance of dying. As with most infectious diseases, children were particularly susceptible. If infected, the fatality rate among children under ten ranged from about 80 to 98 percent.

In the closing years of the eighteenth century, smallpox killed 400,000 Europeans each year. Throughout history, first in the Old World and then later in the New World, smallpox epidemics indiscriminately killed rich and poor alike. With the exception of bubonic plague, smallpox has probably killed more people over the course of history than all other infectious diseases combined.

Survivors were not left unscathed. Imagine a horrendous case of blistering acne all over your body. Where smallpox pustules erupted, they scabbed over. Hardened, crater-shaped pocks disfigured the faces of many survivors. Burst pustules could leave a crust over one's eyes. It's estimated that smallpox caused over a third of the cases of blindness among Europeans in the eighteenth century. So bad are smallpox lesions that they are detectable on Egyptian mummies dating as far back as the fifteenth century B.C. There is even evidence that the young Pharaoh Ramses V died of smallpox several centuries later, in 1145 B.C. And Chinese descriptions of the disease date back to at least the eleventh century B.C.

During these terrible outbreaks, survivors tended to the newly afflicted. It was common knowledge that surviving infection conferred immunity. And while no one knew exactly why or how this worked, they knew it did. In contrast, medical treatment often did more harm than good. Ninth-century Persian physician Al-Razi first distinguished smallpox from measles while chief of the hospital in Baghdad. He recommended that the disease be sweated out of patients to speed the release of "bad humours"—vapors thought to arise from fermentation of the blood.

For centuries, those patients unfortunate enough to afford medical treatment were confined to rooms with blazing fires and sealed windows. European patients endured equally misguided treatments. Leeches were applied to bleed out the fever. Many among the well-to-do experienced the "red treatment." This desperate and illogical practice involved dressing patients in red, and keeping them covered in red blankets in rooms hung with red drapes. The practice continued into the early twentieth century.

Smallpox made its way from Asia to Europe between the fifth and sixth centuries. Trade routes, like the Silk Road that stretched from China to Syria, helped spread the virus around the world. While Europeans had built up some immunity by the time the first explorers reached the Americas in the 1500s, the indigenous people had never encountered the virus before. The experience proved disastrous over the next several centuries. Far more indigenous Americans died from smallpox than in armed battles with European colonizers.

An understanding of how the smallpox virus worked grew in the mid seventeenth century when English physician Sir Thomas Sydenham noted that the highest mortality occurred among those who received the most medical attention. In an early instance of evidence-based medical practice, he went on to reject traditional heat treatments and advocated cooling patients to dispel the fever ravaging their bodies.

Finally, in 1716, Lady Mary Montague, wife of the newly appointed English ambassador to the Ottoman Empire, helped spur a breakthrough after witnessing the Turkish practice of "engrafting" shortly after she'd arrived in Constantinople (modern Istanbul). Bearing the telltale leather-like complexion of a smallpox survivor, Lady Montague was astounded to find that engrafting—as inoculation was called in Turkey—seemed to largely render the disease harmless. Describing the process in a letter to a friend in England, she wrote that the locals held parties for children at which an old woman would make a small cut with a needle and place a pinch of powder, made from the scabs of individuals with the mildest cases of smallpox, into the open wound. (The mild cases were probably caused by a variant of the smallpox virus termed *Variola minor*, which was rarely fatal.) After eight days the children would develop twenty or thirty pocks, come down with a light fever, and recover a few days later, no longer susceptible to the disease.

The practice so impressed Lady Montague that she persuaded the embassy physician, Charles Maitland, to oversee the inoculation of her young son. Maitland watched, aghast, as an elderly Greek woman picked up a rusty needle in her shaking hand and applied dried scab material to a small wound in the boy's arm. Only once her son had recovered did Lady Montague tell her husband what she'd done.

Of course, no one at the time had a clue that immune cells existed, much less that dendritic cells picked up antigen from pathogens and activated adaptive immune cells. But not fully understanding how inoculation conferred immunity didn't stop the practice. It worked, after all. Although it was eye-opening to the English, the counterintuitive practice of inoculation was hardly new. For generations, doctors throughout India had inoculated children against smallpox by puncturing the upper arm with needles dipped in the pus from smallpox pustules. And although the practice only arrived in Constantinople shortly before Lady Montague's visit, the Chinese had already practiced smallpox inoculation for a thousand years. Their approach reportedly involved sometimes blowing a powder of dried smallpox scabs up the noses of children, one nostril for girls and the other for boys. Regardless of method, most of the time children developed, and recovered from, a mild fever and became immune.

Lady Montague and her family returned to England shortly before a smallpox outbreak in 1721. Again she called on Dr. Maitland, who by this time had retired and returned to London himself, to inoculate her young daughter. Maitland agreed, but only if witnesses were present. Physicians of the royal court of King George I obliged. Lady Montague's daughter developed a light fever and suffered only a few pocks.

Soon the king's daughter-in-law, Caroline, the German-born Princess of Wales, heard of this curious practice and was interested in having her two youngest daughters inoculated. Her eldest daughter had contracted smallpox in 1720 and nearly died, and Princess Caroline herself had survived an infection in 1707. She set about collecting information to satisfy both herself and the king, from whom she needed permission. She personally questioned Lady Montague and insisted on seeing her inoculated daughter in person. But the princess remained skeptical and demanded more evidence before daring to try this method on her own children.

She persuaded the king to enlist six inmates from Newgate Prison for an inoculation experiment. In exchange for their participation, they would be granted freedom. All survived, but still it wasn't proof enough for Caroline. The empirically minded princess interviewed doctors and called for orphans from the Parish of St. James to undergo experimental inoculation. Like the prisoners, all the children survived. Finally satisfied, the Princess presented her evidence to the king and secured permission to inoculate his granddaughters.

Not all smallpox inoculations were so successful. Dosages were guessed at and little was asked or known about the underlying health conditions of those inoculated. There was no way to determine if one was inoculating with *Variola major* or *Variola minor.* Some people died after vaccination, which stirred suspicion and fear. And it continued to remain a mystery, even to those administering the vaccines, how deliberately infecting a person could ward off disease or greatly lessen symptoms. Physicians continued to worry about the potential for smallpox inoculation to actually spread the disease. There was also the very real threat of inadvertently passing on other diseases during inoculation, like syphilis, which was rampant at the time.

And, of course, there were those who voiced more chauvinistic and ignoble concerns, like Royal Society Fellow Dr. William Wagstaffe's comments in regard to Lady Montague's experience in Turkey: *"Posterity will scarcely be brought to believe that a method practiced only by a few Ignorant Women, amongst an illiterate and unthinking People should . . . be received into the Royal Palace."*[1]

Despite this sentiment and a general prejudice among physicians, the power of implicit royal endorsement swayed medical opinion, which was further bolstered by compelling statistics. Dr. James Jurin, a scholarly physician with mathematical training, began gathering data on a grand experiment that eventually revealed the mortality rate from smallpox had dropped from around 20 percent among the uninoculated to less than 2 percent among the inoculated. Inoculations also greatly lessened the disfigurement and blindness that so often afflicted smallpox survivors.

Lifesaving inoculation practices arrived in Boston during the smallpox epidemic of 1721. It was not the first time the disease had struck New England, but it was among the worst. Approximately half of Boston's 12,000 residents became infected and almost 900 died. Puritan minister Cotton Mather was deeply concerned about the outbreak and decided to act. He was aware of Italian inoculation practices and learned about African inoculation practices from Onesimus, a slave he had recently acquired. Mather enlisted the services of a physician, Zabdiel Boylston, who used his own son and two slaves as test subjects. All survived, and inoculation of almost a fifth of Bostonians soon followed in the first-ever vaccination campaign against infectious disease in America.

Though an undeniable achievement, it was not universally welcomed in colonial society. Many traditional medical and religious groups raised intense opposition to the vaccination campaign, outraged over such an affront to the will of God. Mather's house was bombed as the epidemic raged. Undeterred, Mather and Boylston collected information on fatalities. They pointed to a 14 percent fatality rate among those who became naturally infected with smallpox, compared to a 2 percent fatality rate for those who received inoculation. Such information held little sway with the naysayers, but convinced Mather that vaccination was a gift from God.

While effective, inoculation practices did border on the barbaric at times, no doubt contributing to public anxiety over the new practice. Accounts are rare, but in 1757 a curate (priest's assistant) in a Gloucestershire parish recorded the ordeal of an eight-year-old orphan named Edward Jenner. Apparently, doctors made quite a show of it, putting the poor lad through a series of fasts and bleedings for six weeks before inoculation. When the time arrived, the child was hauled into a stable, starved and anemic. Reportedly haltered like a horse, he underwent the quick inoculation procedure. Decades later, Jenner would go on to make a remarkable contribution to the development of a smallpox vaccine.

Orphaned as a young child, Jenner was looked after by his older siblings. The boy readily took to the countryside around the town of Berkeley in the Severn Vale, about a hundred miles east of London. He had an intense interest in natural history and studied the plants and animals in the vicinity of his home and school. Later, having no means to obtain an Oxford education, Jenner apprenticed to a local surgeon. He kept up with his naturalist studies and caught the eye of a famous botanist, Joseph Banks, who had accompanied Captain Cook on his first scientific voyage. Banks needed someone to catalog botanical specimens and other natural treasures brought back from the voyage. Still in his early twenties, Jenner jumped at the chance.

He completed the task but declined an invitation to go on Captain Cook's next voyage, preferring to remain in England and continue his medical and naturalist studies. Over the next two decades Jenner became a successful country doctor. He indulged his curiosity in the natural world through the study of cuckoo birds. All cuckoos lay their eggs in the nests of other birds, and Jenner discovered some odd goings-on in such nests. He observed cuckoo fledglings throwing the other eggs overboard, or, on other occasions, throwing out the fledglings of the parent birds. After investigating the phenomenon further, he concluded that the odd behavior was actually quite common. In 1788, he wrote up his findings and presented them to the Royal Society for publication. Not all of the Society's Fellows believed Jenner's discoveries of outlandish cuckoo behavior. How could a bird could do such a thing? Nonetheless, with the help of influential allies like Joseph Banks, Jenner's work on cuckoos was published, securing his place as a Royal Society Fellow at the age of forty.

But Jenner was just as interested in the world of medicine as he was in the bizarre behavior of cuckoos. And, as smallpox outbreaks continued in London into the early nineteenth century, doctors responded with sporadic inoculation campaigns. At the same time, folk stories were circulating that milkmaids afflicted with cowpox never seemed to come down with smallpox. Cowpox was not a fatal disease, but it was still bothersome to farmers, milkmaids, and, no doubt, cows. Whenever an outbreak of cowpox occurred, cattle developed blisters on their udders and the milkmaids who milked the infected cows developed blisters on their hands.

As a country doctor tasked with vaccinating his patients during outbreaks, Jenner kept up to date with the techniques and results of smallpox inoculation practices in London. Milkmaids were among his patients, and he became intrigued by their response to inoculation. True to the stories, they never developed the slight fever or the few pocks that others did. His skill at observing and recording the natural world proved handy when he noticed an important detail—a small blister always seemed to form at the inoculation site but nowhere else.

When the next cowpox outbreak occurred, he recruited a milkmaid with fresh cowpox lesions and a young boy who had never had smallpox nor been inoculated. On May 14, 1796, he extracted fluid from a small blister on the infected milkmaid's hand with a lancet. He quickly inserted the fluid-filled lancet into the boy's arm. The boy suffered a headache, loss of appetite, and mild malaise for a day and a night.

A couple months later, Jenner collected pus from someone with a virulent case of smallpox and again inoculated the boy. The lad had no reaction this time. Jenner's suspicions were confirmed; he could use cowpox to prevent a person from contracting smallpox. During the next cowpox outbreak, in 1798, Jenner performed similar experiments and concluded that inoculation with the mildly pathogenic cowpox protected against its deadly cousin, smallpox.

In addition to cows and milkmaids with active cowpox infections, doctors could now use material from pustules of those newly inoculated with cowpox as a ready source of smallpox inoculant. Using cowpox rather than smallpox as inoculant also extinguished the possibility of spreading smallpox or dying from an inoculation of *Variola major*. The practice spread rapidly across Europe.

Jenner had discovered how to use the barely virulent cowpox on people to protect against highly virulent smallpox. Here lay the key to successful vaccines: stripping a pathogen of its virulence, while retaining the aspects that trigger an immune response. The secret behind Jenner's success, we now know, was stimulating adaptive immune cells, the ones with the ability to remember and recognize a specific molecular signature of pathogens after the first encounter. Jenner never faced the dilemma that polarized Sabin and Salk because cowpox simply couldn't kill people. Still, not everyone was impressed with his achievement.

In 1797, Jenner sent a report of his observations and experiments to the Royal Society for publication. Despite being a Fellow himself, and with the strong recommendation of the famous botanist Joseph Banks, then president of the Royal Society, the society rejected Jenner's outlandish report. Undeterred, Jenner self-published his seventy-five-page report a year later. Interestingly, the report does not use the word "vaccination," though Jenner's research is certainly the inspiration for the word. It was a surgeon friend of Jenner's who coined the term from *vacca*, Latin for "cow."

Annual statistics on smallpox fatalities after the practice of inoculation was in place reveal that deaths dropped by a factor of six from 1801 to 1875. In Sweden, where vaccination was compulsory after 1816, the annual death toll from smallpox dropped a thousandfold, from 12,000 in 1801 to just eleven in 1822.

Jenner had pulled off a medical coup that would eventually vanquish smallpox. He also placed another stone on the path to germ theory. It took two centuries after Jenner's first demonstration of a smallpox vaccine for the disease to be eradicated. The world's last natural case occurred in Somalia in 1997. Today, the World Health Organization is debating what to do with stockpiles of the virus stored in government labs since the height of the Cold War.

ON THEIR HANDS

Like the practice of vaccination, new hygienic standards helped reshape medicine. In the late 1840s, decades before scientists realized that microbes caused disease, Hungarian physician Ignaz Semmelweis promoted the then-radical concept of hand-washing. At the time, he worked in the maternity ward at Vienna General Hospital, where many women giving birth contracted childbed (or puerperal) fever, an infection of the reproductive tract. As many as one in three women died from the mysterious ailment in some maternity wards. Semmelweis made the impolitic observation that his hospital's training ward for doctors had three times the mortality rate as the ward used to train midwives. The key difference was that doctors wore their bloody coats with honor and went from patient to patient, and from autopsies to patients, without washing their hands. Semmelweis began insisting that doctors change their coats and clean their hands with chlorinated lime (calcium hypochlorite) after autopsies and before seeing live patients. These simple measures reduced mortality by 90 percent in the ward staffed by doctors—down to the level of the ward staffed by midwives.

His success infuriated the medical community. By connecting rampant puerperal fever to poor hygiene, he not only incriminated doctors but ran afoul of the mainstream medical understanding that disease was caused by "bad air"—the miasma theory of the ancients. Despite his demonstration that hygienic practices did actually work, Semmelweis could not explain exactly *how* they worked. Of course, doctors did not take kindly to the suggestion that they wash their hands either. How dare he imply that the hands of gentlemen such as themselves were unclean. Or that female midwives were doing a better job.

Shunned by colleagues, Semmelweis was summarily dismissed from his position at Vienna General. He moved to Budapest and assumed the unpaid post of honorary head physician in the maternity ward of a small hospital where childbed fever was rampant. He immediately instituted hand-washing and once again virtually eliminated the disease. He was met with the same response by his new Hungarian colleagues: they scoffed at the practice and summarily rejected the ludicrous notion that washing one's hands could prevent the spread of disease. Ceaseless, blistering criticism took a heavy toll on the poor doctor. Suffering from severe depression, he died in a mental institution just as pioneering microbiologists began to provide conclusive proof that microorganisms caused human diseases. Today, philosophers describe out-of-hand rejection of new knowledge that contradicts established beliefs or paradigms as the "Semmelweis reflex."

10

FEUDING SAVIORS

The man who saved France's wine and vinegar industries with his discoveries of the secret power of bacteria also helped revolutionize the medical view of microbes. In 1865, the year a broken Ignaz Semmelweis died, Louis Pasteur was once again called away from his laboratory to address a practical problem. The silkworm industry was suffering from a mysterious malady that was decimating the economy of southern France. When his aging mentor, the famous chemist J. B. Dumas, begged Pasteur to look into the cause of the silkworm die-off, the younger chemist reluctantly agreed, cautioning that he didn't know the first thing about silkworms, or any other worm for that matter.

Regardless, Pasteur agreed to investigate the silkworm disaster out of loyalty to his old professor. When he arrived to help, the growers showed him sick silkworms covered with little black specks. Pasteur delicately cut open a few of the silkworms and noticed microscopic globules in the fatty tissue of their bellies that he immediately suspected were sure signs of disease. This looked like a simple problem to solve. Pasteur told the committee of farmers that they should use a microscope to examine each pair of moths at mating time and use only the eggs from those that had no such globules in their bellies.

Resistant to changing their age-old practices, the farmers protested. They claimed that they could not possibly use such a contraption. Pasteur countered that if an eight-year-old girl in his laboratory could master the microscope, then surely they could. This was enough to shame the skeptical growers into buying one. The next breeding season, they carefully collected eggs from moths with globule-free bellies, and watched like hawks when the precious eggs hatched into worms the next spring. Pasteur's confidence—and reputation—crumbled as sickly worms emerged from the eggs.

Facing bitter mockery from desperate silkworm growers, Pasteur examined more worms and found that some sickly broods had no globules in their belly fat. Perplexed, he fed leaves smeared with the excrement of sick worms to a batch of healthy worms. They all died. This little experiment provided the final piece of the puzzle: The tiny globules were indeed microscopic parasites, but they didn't necessarily roost in a silkworm's belly fat. The way to protect the silk industry was to keep the healthy worms away from leaves that had been soiled by the sick ones.

The growers reluctantly followed his advice a second time, and were delighted to find that it worked. Healthy worms emerged from the eggs and Pasteur was lauded as the savior of the silk industry. This experience taught Pasteur an important lesson. Microbes could do more than make or degrade beer and wine—they could convey disease to animals. From there it wasn't much of a leap to extend his investigations of microbes into diseases that affected people.

Pasteur went on to prove that he could prevent mold from growing if he excluded dust particles from a flask of boiled broth. This result didn't just further discredit the long-held, stubbornly embraced theory of spontaneous generation; it proved that the growth of mold required seeding from "germs" in the air. Pasteur's demonstration that microbes did not arise from putrefying matter inspired medical researchers to investigate how to prevent bacterial infections in the first place.

Through his work with the wine and vinegar industries, Pasteur had transformed the perception of microbes from van Leeuwenhoek's curious playthings to mankind's invisible helpers. In solving the mystery of what sickened silkworms, Pasteur increasingly turned his studies toward investigating the role that microbes play in causing disease. Naturally enough, this kind of thinking fostered the perception of microbes as merciless killers and the idea that germs warranted ruthless extermination. Pasteur yearned to wipe infectious diseases off the face of the globe, a high-minded vision that landed him in a bitter feud with a stubborn young Prussian doctor named Robert Koch.

As a boy, Koch dreamed of exploring remote corners of the world. But, when he graduated with a medical degree from the University of Göttingen in 1866, the only work he could find was as an intern at a Hamburg insane asylum. This stint was followed by a succession of uninspiring medical practices in a series of disease-wracked Prussian villages, all of which conspired to keep Koch isolated from the rapidly advancing world of microbiology. Until, that is, his wife bought him a microscope to take his mind off his tedious practice, not realizing at the time how important this gift would be to the world.

Koch began aimlessly using his microscope in much the same way that Leeuwenhoek had—indulging his natural curiosity by examining everything he could think of. One day he placed some blood from sheep and cows that had died of anthrax under his microscope and noticed tiny quivering things that looked like little sticks. Were they alive? Were these the microbes that M. Pasteur was raving about? His suspicions were aroused when he examined the blood of healthy livestock and found none of these little living rods.

By the following year, he observed that the rods from these same blood samples had changed into spores. Here was the explanation for the well-known mystery of how anthrax could afflict sheep that grazed in certain "sick" fields. The spore life phase of the pathogenic microbe could survive in the soil until a new host came along. This was why sheep could contract anthrax from soil as well as from diseased animals.

Working at home in a primitive laboratory, Koch cultured the blood of anthrax-infected mice and in 1876, still in his early thirties, he published a paper on the bacterial origin of anthrax. The following year Koch isolated the bacterium responsible for anthrax through experiments that, for the first time, definitively linked a specific disease to a specific microorganism. Koch not only discovered this crucial link, but he suggested a simple and effective way to control the disease—burn the bodies of animals that died from it instead of burying them. This practice, he contended, would prove to block transmission through bypassing the spore phase of the bacterium's life cycle. The European scientific community hailed a new rising star.

Koch was an ingenious researcher. He stained bacteria with different-colored dyes to make them easier to see and differentiate. Convinced that confusion over descriptions of bacteria would hinder progress in identifying pathogens, he bought a camera. Placing its lens up against his microscope, Koch took the first photographs of bacteria.

Sure that every disease had its own distinct microbe, Koch wanted to grow pure cultures of single types of bacteria. But how? The liquid soups that microbiologists used to grow microbes were prone to contamination from airborne microbes.

Koch found the key one fateful day after noticing that half a boiled potato accidentally left on his laboratory table was covered in a rainbow of tiny colored dots—gray, red, yellow, and violet. Intrigued, Koch scraped a tiny sample from each dot and examined it under his microscope. The dots were different bacterial colonies—one held round-shaped bacteria, another tiny rods, and yet another was comprised of living corkscrews.

Here was the answer for how to grow pure cultures of microbes: plant a microbe on a surface made of nutrients, cover it to prevent contamination, and let the microbes grow into a colony. Experimenting further, he mixed gelatin with beef broth. The microbes loved it. Finally, he had a way to grow pure colonies of suspected pathogens.

Koch, the physician, brought a diagnostic rigor to establishing the causal relationships between specific diseases and particular microorganisms. In so doing, he developed an unwavering belief in the fixed nature of microbes and their specificity in causing disease. He also believed sanitation was the way to control disease.

Although Koch and Pasteur used different names for the culprit bacterium that caused anthrax, their mutually supportive evidence confirmed that a microbe lay at the root of the disease. Their work solidly established the germ theory of disease. However, the two took radically different approaches to applying their science to combat disease.

Pasteur, the experimental chemist, excelled at solving practical problems by synthesizing observations into novel practices. Zeroing in on the natural variability of bacteria, he suspected that using weakened or attenuated varieties could make effective vaccines. With this idea in mind, Pasteur shifted his research to the prospect of using microbes to cure the very diseases they caused.

Aware of Koch's success in isolating the bacterium that caused anthrax, Pasteur set out to save the sheep of France from this deadly and highly contagious disease. Every year since ancient times it had mysteriously killed thousands of people and far more sheep. One day a shepherd might check on his flock and conclude all was well. Rising the next morning, he'd look across the field to find his sheep keeled over, dead for no apparent reason.

After long weeks of experimentation Pasteur found that heating the anthrax-causing bacteria to high enough temperatures produced strains sufficiently attenuated to vaccinate livestock. In late January 1881, one of the editors of *Veterinary Press* penned a skeptical article about Pasteur's experiments. Two months later, the same editor called for a field trial of Pasteur's laboratory studies on anthrax. Would they work on a real farm? Pasteur accepted the challenge and put his idea to the test on sixty sheep provided by the Melun Agricultural Society.

Twenty-five healthy sheep were inoculated with an attenuated strain of anthrax that could kill guinea pigs but not larger animals. Two weeks passed and the sheep were inoculated again, this time with attenuated anthrax strong enough to kill a rabbit. For the final phase, a few days later, Pasteur exposed the original twenty-five sheep, plus twenty-five unvaccinated sheep, to the most virulent anthrax strain in his laboratory. A month after the experiment began, all the unvaccinated sheep lay dead. The vaccinated sheep were all healthy and thriving. The dramatic success of this field trial silenced Pasteur's critics and detractors—at least in France.

That summer both Pasteur and Koch attended the Seventh International Medical Congress in London. Pasteur presented a paper on his recent anthrax vaccine field trial. Koch talked about his new method for staining and identifying bacteria. The famous Frenchman pointedly praised the younger German researcher, calling his work a major advance. It was the first—and last—friendly exchange between the two. Koch simply didn't believe in Pasteur's attenuated microbes. He was convinced that the results of Pasteur's experiment must have another explanation.

Just months after the London meeting, Koch and two of his students published a paper challenging Pasteur's claim that he had created attenuated strains of anthrax. They accused the Frenchman of using impure cultures, misidentifying the microbial culprit, and botching the inoculation study. All adding up, they asserted, to sheep that hadn't experienced a full-blown anthrax infection, and therefore had not been cured of anything! In September 1882, several months after Koch's own fame skyrocketed after discovering and isolating the bacterium responsible for tuberculosis, the two met again at a conference in Geneva. In his presentation Pasteur addressed Koch's criticism of the anthrax vaccine field trial.

Neither man spoke the other's language, so a colleague sitting next to Koch rapidly translated Pasteur's words into German. *Too* rapidly it turned out. When Pasteur referred to Koch's published work as "*recueil allemande*," a collection of German works, Koch's translator rendered the phrase as "*orgueil allemande*," German pride or arrogance. After Pasteur's speech, Koch rose and angrily vowed to respond in writing. Pasteur was perplexed by his colleague's reaction. As promised, Koch followed up the exchange with another paper, this time impugning Pasteur's work on anthrax inoculation as useless, going so far as to question why physicians should listen to a chemist with no medical training.

Undaunted by Koch's attacks on his anthrax work, Pasteur turned to another dreaded disease, nearly always fatal and quite common at the time—rabies. He experimented on rabid dogs and found that inoculating them with weakened strains prevented the development of serious symptoms. One shudders at the thought of a lab filled with rabid dogs and the details of Pasteur's assistants drilling into their brains to collect rabies virus. Nevertheless, these dangerous and grueling experiments had a big payoff.

On Monday, July 6, 1885, nine-year-old Joseph Meister and his mother arrived on the lab's doorstep. Two days earlier, on his way to school, a rabid dog had attacked young Meister. His thighs, legs, and hands were a mess of dog-bitten flesh. This was every parent's worst nightmare—the bite of a rabid dog was a death sentence. Pasteur's successful experiments on dogs had been reported in the press, but experimenting on a person was another thing. After consulting with several physicians, who all agreed Meister had no chance of survival, Pasteur inoculated the boy with attenuated rabies vaccine twelve times over the course of ten days. On the last day, he inoculated Meister with a freshly collected, highly virulent strain to test whether the treatment had indeed conferred immunity.

The boy survived the rabies and, amazingly, infection stemming from the dog-bite wounds didn't take young Meister down either. The wounds healed and in two months the boy returned to perfect health. As soon as Pasteur's paper describing his experiment on Meister was published, desperate dog-bite victims began showing up at Pasteur's lab. Soon the first order of each day's business was treating new patients, including those who had come from as far away as New York City. Pasteur was now an international celebrity and a scientific superstar—a miracle worker curing diseases that had long haunted humanity.

THE ROOTS OF GERM THEORY

Though the two rivals took fundamentally different stances on microbes themselves, Koch and Pasteur are considered the founders of medical microbiology. Under Koch's influence, German microbiologists developed standardized methods for studying bacteria that paid off in discovering the causes of many diseases. Although there are cases in which several pathogens underlie ailments, the idea of one microbe per disease worked pretty well.

Koch developed a rigid view that the specific properties of microbes were fixed and immutable. He didn't believe virulence could vary, rejecting the principle at the heart of Pasteur's attempts to develop vaccinations from weakened pathogens.

In contrast, Pasteur believed that changes in virulence could help explain sudden outbreaks of infectious diseases. He reasoned that microbial pathogens could change over time and even jump to new hosts. Under Pasteur's influence the French school focused on immunity and vaccine development.

Koch's discoveries led in a different direction. His approach required isolating pure bacterial cultures in order to identify and study disease-causing organisms. He set down four postulates that underpin germ theory and are still used today: (1) the microbe must always be found in the body of those with the disease; (2) it must be isolated from the host and grown in pure culture; (3) the same disease must be caused when the cultured microbes are reintroduced into a susceptible host; and (4) microbes retrieved from a deliberately infected host must match those present in the original sample. When one can demonstrate all four of these conditions, one can confidently confirm a causal relationship between microbe and disease.

By the end of the nineteenth century, the isolation of microbes responsible for gonorrhea, leprosy, plague, pneumonia, syphilis, tetanus, and typhoid erased the skepticism of the last opponents of germ theory. In 1905, Koch was awarded a Nobel Prize for his work on tuberculosis, which, prior to his discoveries, had been widely considered an inherited disease.

Given Koch's unrelenting focus on identifying pathogens underlying disease, and that, unlike Pasteur, he had little experience with beneficial microbes, one wonders if he had any inkling that he was culturing only a mere sliver of the microbial world. We are fortunate, of course, that pathogenic microbes garnered his attention, but such an intense focus on those that harm us fostered a one-sided view of microbes that has been hard to shake. Widespread adoption of Koch's postulates constrained research in the field of microbiology to culturable organisms. A microbe that could not be grown in a culture could not be isolated, and therefore could not be studied using Koch's postulates—so they weren't.

Today, microbiologists know that only a tiny fraction of microbes are easily and reliably cultured. While culturable microbes cause many of the deadliest diseases, the general acceptance of germ theory inhibited scientists from exploring and researching microbes in their ecological context. The goal throughout much of the past century was not so much to understand the microbial world, but to isolate and destroy particular pathogenic microbes.

The combined work of Pasteur and Koch left no doubt about the validity of germ theory. The idea that tiny, invisible creatures could enter our bodies and cause diseases was a radical departure from long-standing conventional wisdom dating back to the age of Classical Greece. Until this feuding pair demonstrated otherwise, physicians didn't know if bacteria and other microbes were the cause or the result of disease.

Toward the end of the nineteenth century, germ theory became a foundational concept in medicine in the way that Darwin's theory of evolution influenced biological thought. The terrifying prospect that an invisible speck could take a person down united humanity against a common foe. Once we saw the harm and horror that microbes could cause, we saw them all that way—a view that still colors the way most people see microbes today.

And yet, despite the revolutionary work of Robert Koch and Louis Pasteur, a mystery remained. There was a group of serious ailments that seemed to defy germ theory. No bacteria were to be found at the root of rabies, measles, smallpox, influenza, and a handful of other seemingly intractable diseases. It was a puzzling and frustrating conundrum until, in the early twentieth century, experiments established that a class of infectious agents too small to be seen with an ordinary microscope existed. They could pass through filters that trapped bacteria.

The invention of the electron microscope in 1931, with its exquisite powers of magnification, helped solve the mystery. The culprits were small indeed—and not live bacteria at all, but the not-quite-alive viruses. Although they occupied a special place on the living world's inanimate frontier, viruses joined the club of microbes that defined and sustained germ theory.

THE WONDER DRUGS

Pasteur's vaccines were a marvel, but they were not enough to vanquish disease. They, in fact, didn't and couldn't kill a microbe. They could only impart immunity. And with microbes singled out as the enemy, scientists from different fields concentrated on finding ways to stomp them out. Nature herself would prove to hold a treasure trove of murderous agents.

In 1928, a famously disorganized and brilliant Scottish physician rushed away from his perpetually messy lab at London's St. Mary's Hospital to start a long vacation and left some bacteria-culture plates laying around. When Alexander Fleming returned, he discovered a wonder drug had sailed into his lab through an open window. The culture plates were covered in mold. As he was cleaning up the fuzzy mess, he noticed that some of the plates had a zone clear of bacteria around the fungal colonies. Somehow this fungus inhibited the growth of the bacteria. Fleming cultured the uninvited guest (*Penicillium notatum*) and isolated its antibacterial compound. He named it penicillin.

Fleming's previous experience in the First World War fueled his interest in researching antibiotics. He served as an army surgeon in battlefield hospitals and helplessly watched thousands of young men die of infections stemming from nonfatal wounds. Although highly motivated to find antibacterial agents, even Fleming initially overlooked the importance of his discovery. He was occupied on other fronts, having gained notoriety for his discovery of a substance with antibacterial properties in the nose of a patient with a head cold. Innate immune cells created the substance, which Fleming named lysozyme. This naturally produced antibacterial was later found in other bodily fluids, including saliva, tears, breast milk, and mucus. If our bodies and a lowly fungus could produce compounds with the power to kill bacteria, then surely other substances lay waiting in nature's pharmacy.

Although he promptly published his discovery of penicillin in 1929, the paper attracted little interest. Fleming went on to successfully use penicillin to treat a few patients with eye infections, but he could never culture enough of his novel fungi to pursue clinical trials. The good doctor moved on to other projects, and the power of penicillin would go undiscovered for nearly a decade, until colleagues at Oxford developed a way to grow the mold. In tests on mice that followed, penicillin proved incredibly effective. The next step was a clinical trial on a small number of patients in 1941. The new drug performed flawlessly, saving two people from certain death. Though clinical trials were meager, the results were good enough. Mass production of penicillin began in earnest and supplies were rushed to the front lines as another world war raged across Europe. Anything that could curb the death of troops from infection and control another major military malady, gonorrhea, would be a powerful weapon the other side didn't have.

In the years between the two world wars, industrial chemistry had grown into a burgeoning new field. It was also proving to be an interesting and productive place to look for antibacterials. Four years after Fleming's accidental discovery of penicillin, a researcher at Bayer laboratories in Germany was investigating possible medical applications of industrial dyes. Gerhart Domagk discovered that a textile dye, prontosil rubrum, cured mice infected with *Streptococcus* bacteria. Human trials got under way and the dye proved effective at killing bacteria in humans as well. Unfortunately, it also caused kidney damage and turned the skin a striking shade of bright red. Domagk wasn't all that taken with the results. Nonetheless, the dye was patented for use as a drug in 1934 under the trade name Prontosil.

A year later Domagk's six-year-old daughter, Hildegarde, fell down the stairs at their home in early December 1935. This shouldn't have been a near-death experience but the girl happened to be carrying a sewing needle. She had been making a Christmas ornament and wanted her mother's help threading the needle. A good part of the needle pierced her hand, eyelet-end first, and broke off. Little could be done at home so Domagk rushed her to a doctor. The needle fragment was extracted and Domagk and Hildegard returned home glad to have the accident behind them. Or so they thought.

A few days later her hand started swelling and an abscess formed at the site of the wound. A streptococcal infection had set in. The doctors opened and drained the abscess three times. Domagk became increasingly concerned, well aware that uncontrolled infections could kill a person. He waited a few more days and panicked when he saw red streaks flying up Hildegard's arm accompanied by a fever. As the girl's condition worsened, the doctor told Domagk that Hildegarde faced amputation. A frantic Domagk dashed off to his lab and returned to the hospital with a supply of Prontosil tablets. It was still an experimental drug, but over several days he gave her increasing doses scaled up from what he'd used to cure the mice in his lab. Finally, it did the trick. Hildegarde miraculously recovered and was home by the holidays. Prontosil had saved his daughter.

Soon other researchers, investigating the mechanisms by which Prontosil proved lethal to bacteria, discovered that a particular portion of the dye molecule killed bacteria. This discovery led to the development of the first commercial group of antibiotics, the sulfonamides. Despite the earlier discovery of penicillin, sulfa drugs beat it to market in 1937. At last the prospect of killing disease-causing bacteria was more than a glimmer in the eyes of scientists.

In 1939, Domagk was awarded a Nobel Prize only to run afoul of Adolf Hitler. Several years earlier an anti-Nazi pacifist, Carl von Ossietzky, received the Nobel Peace Prize for exposing Germany's clandestine rearmament, an act that incensed the rising dictator. In retaliation, Hitler decreed that no German could receive a Nobel. Domagk danced around the decree, writing the Nobel organization of his gratitude for being selected. Unamused, the Gestapo arrested Domagk for being "too polite to the Swedes" and held him for a week before persuading him to sign a letter declining the award. In 1947, the war over and Germany in ruins, Domagk was finally able to claim his gold medal. But he never received the prize money; it had been redistributed in accordance with the rules of the Nobel Foundation.

Even as advances based on germ theory promised triumph over humanity's ancient microbial foes, the seeds of a different view had already begun to take root in a backwater of soil science. At the start of the twentieth century, a young immigrant to the United States fell for a certain group of soil-dwelling bacteria. Selman Waksman grew up in a rural village in western Ukraine. As a Jew, he knew he'd never be admitted to the University of Odessa on the shores of the Black Sea so, in 1910 at the age of twenty-two, he left his homeland to pursue higher education.

When Waksman arrived in the United States, he stayed with a cousin and her husband on their farm in rural New Jersey near what is now Rutgers University. Although he was admitted to medical school at Columbia, his wide-ranging interests took him elsewhere. Working on his cousin's farm sparked an interest in soil and how compost could improve soil fertility. Waksman's new interest so intrigued him that he opted to study agriculture instead of medicine.

In 1912, he won a scholarship to Rutgers and flourished under the tutelage of a professor who steered Waksman into soil microbiology. Medical researchers of the day, thoroughly enamored with germ theory, had a near-exclusive focus on how to control and eradicate human pathogens. But in agriculture there was an emerging awareness of and interest in the great diversity of life forms that lived in soil. As it would turn out, both fields would come to greatly influence Waksman's career.

Waksman needed to do a "practical project" to fulfill his graduation requirements. He settled on culturing the bacteria and fungi from soil samples from the Rutgers farm. A particular group of bacteria caught his eye. They had a leathery texture and conical shape. Sometimes they grew as vibrant blue colonies. While this group of bacteria captivated him, no one else seemed much interested. His professors told him only that such bacteria were generally designated as actinomycetes.

Today, we know that the "earthy" smell of soil is often attributed to this group of bacteria. The actinomycetes are one of the primary groups that decompose organic matter in the soil. While there is no formal definition for an "earthy" smell, it appears that the metabolites that actinomycetes generate in the decomposition process are as distinctive as those of the bacteria used to make particularly odoriferous cheeses.

Waksman must have appreciated the scent that actinomycetes lent to soil because his interest in these strange bacteria only grew. He went on to get his Ph.D. in soil microbiology at the University of California, Berkeley, before returning to New Jersey in 1918, just as World War I ended. Although a former mentor created a position for him as a microbiologist for the Rutgers Farm, there was little in the way of a salary. Waksman worked one day a week at the farm and the rest of the work week at the Takamine drug company researching a newly discovered drug called Salvarsan, which killed the bacteria that caused syphilis. Though a breakthrough drug, Salvarsan was also quite toxic, as it came from an arsenic-based dye. It was Waksman's job to test the toxicity of Salvarsan on human cells.

By the early 1920s, the economy had begun to turn around and Rutgers offered Waksman a position as an assistant professor. He left his job at Takamine and dedicated himself to studying actinomycetes. His laboratory facilities were in shambles, nothing like those of the commercial lab he'd just left. With no graduate students and few assistants, Waksman made do with what he had, churning out the first textbook on soil microbiology.

Had two important scientific events not occurred in the years that followed, Waksman might well have remained focused on actinomycetes in the soil. But the year after his textbook came out, Fleming discovered penicillin. Although derived from a fungus and not a bacterium, it too hailed from nature. The second event stemmed from the work of René Dubos, an energetic Frenchman who became one of Waksman's graduate students in the mid-1920s. Dubos studied how bacteria broke down cellulose, one of the most resistant of all plant tissues to bacterial enzymes. It was the perfect project for Dubos's next endeavor at the Rockefeller Institute, where he searched for substances that could destroy the protective polysaccharide coat around a strain of bacteria that caused pneumonia.

At the time, it still wasn't fully understood why, in the case of some diseases, you could bury dead people or animals in the soil and later find only a few or no pathogens. Was it due to an unsuitable environment? Or, could the microbes that lived in soil kill the introduced pathogens? Lorenz Hiltner and Sir Albert Howard, whom we met earlier, had found this to be largely true in the botanical world. They'd seen how pathogens fared worst in soils chock-full of nonpathogenic microbes. Soon, Waksman and others would look to the soil for compounds to use in combating human pathogens.

In the context of these discoveries, Waksman realized that his foundation in soil microbiology put him in a unique position to build a bridge between the worlds of soil and medicine. Could compounds that soil microbes produced be harnessed for medical purposes? Would his hunting expedition bear fruit, or turn into a futile needle-in-a-haystack endeavor?

He and his graduate students initially began looking broadly at fungi and bacteria, including the actinomycetes. Based on preliminary experiments, Waksman's team quickly cast all but the actinomycetes aside. In 1940, one of Waksman's graduate students found an intriguing compound. They named it actinomycin. But looking promising is not the same as being useful. Further experiments confirmed that actinomycin had too much firepower. It killed lab animals as readily as it killed pathogenic bacteria.

Three years later, the breakthrough finally came. Toiling away in a basement lab, another of Waksman's students, Albert Schatz, found an actinomycete (*Streptomyces griseus*) that produced a compound that didn't harm lab animals and handily killed pathogens. He and Waksman named the substance streptomycin. Schatz painstakingly conducted further experiments and found that streptomycin could kill one of the most serious scourges of all time—tuberculosis.

In 1944, Waksman published the findings about streptomycin, including Schatz as a coauthor. Soon after, the drug company Merck started human trials with the Mayo Clinic to test streptomycin on people with tuberculosis. By the end of 1946, the trials were completed and the results astonishing—streptomycin could cure tuberculosis. By mid-1947, Merck and other drug companies were producing about 1,000 kg of streptomycin every month. In the decade that followed, Waksman's lab continued to comb the soil for antibacterial compounds. It proved quite productive—they found ten more antibiotics.

Of all the compounds Waksman isolated from the actinomycetes, streptomycin proved the most profitable and the most effective over the longest period of time, saving the lives of countless people. Much as the polio vaccine made Jonas Salk a household name, streptomycin put Selman Waksman's name in the headlines. In 1952, he received a Nobel Prize for his work on streptomycin.

Almost immediately, Waksman's former student, Albert Schatz, created a stir when he challenged his advisor's Nobel. For who had truly discovered this marvelous product of the earth? The man with a lifelong love of actinomycetes and their ability to imbue soil with fertility and its telltale "earthy" smell, or the dedicated man who sifted through nature's offerings in the right place at the right time in the other's lab?

The flood of antibacterial compounds from soil bacteria in Waksman's lab, along with the earlier discoveries of penicillin and the sulfa drugs, created a postwar gold rush for new antibiotics. By the 1960s hundreds of new antibiotics had been discovered, and an extraordinary range of formerly serious bacterial infections and diseases could be cleared up by swallowing a couple pills every day for a week or two.

Not only could lives be saved, but money could be made. Antibiotics were as close as one could get to a perfect product. In the span of only a decade or two they seemingly ensured that Americans would never again live beneath the cloud of fatal infections and epidemics. Armed with this powerful new weapon, we declared impending victory in the war against microbes. But in our hubris we overlooked a chink in our antibiotic armor. On December 28, 1940, just before the start of the mass production of penicillin, the journal *Nature* published a prescient paper. The authors, one of whom was the biochemist who helped transform Fleming's mold into a drug, had some troubling news. A bacterium called *B. coli* (later renamed *E. coli*) had conjured up an enzyme that could break down penicillin when exposed to it in a laboratory experiment.

THE PRICE OF MIRACLES

By the end of the decade, researchers ran into another problem. Streptomycin stopped working for some tuberculosis patients. Indeed, almost all the antibiotics Waksman's lab discovered, as well as many others, soon triggered resistance in their bacterial targets. Later it was discovered that bacteria had ingenious countermechanisms. For example, certain bacteria could turn on the equivalent of a high-powered sump pump to purge themselves of antibiotics. Other bacteria could make compounds that shredded an antibiotic into useless pieces. Or, a bacterium could shape-shift, changing a structural protein to thwart the attachment of the antibiotic and thereby cheat death.

Despite such worrying signs, antibiotics were so obviously useful that caution seemed pointless. Why fight progress? The postwar years ushered in the era of modern chemistry that promised solutions, not problems. The kill-them-all philosophy behind agricultural pesticides and herbicides stormed through the medical world as well. With every passing year more antibiotics were made, and more pathogens pushed back. And we've been trapped in this cycle ever since.

Still, it's likely that antibiotics have saved at least one member of every family in America today. When you need them, they are truly miraculous. And yet, overlooked in the rush to embrace these truly wondrous drugs was the evolutionary implication of a twenty-minute lifespan. An antibiotic never kills all the bacteria at the root of an infection. The bacteria that survive an antibiotic course go on to reproduce as their mates disintegrate around them. Most important, the survivors pass on the genes that conferred the trait(s) that allowed them to successfully dodge the antibiotic. This simple reality is the Achilles' heel of antibiotics.[1]

For the past half-century antibiotics have been routinely overprescribed, leading to increasingly drug-resistant bacteria. Few, however, know about a far more troubling, ongoing abuse of antibiotics—their massive application to healthy livestock to stimulate growth. Animals fatten up sooner on antibiotics than they would otherwise. Globally, about 90 percent of all antibiotics used are given to animals with no apparent infection. This is an even more effective recipe for promoting development of drug-resistant bacteria, and it's working.

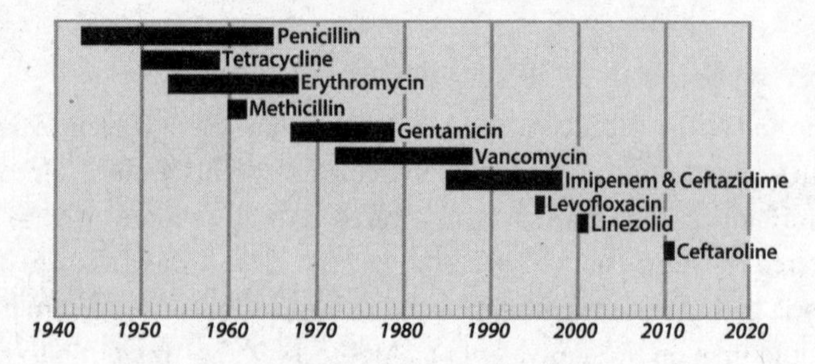

Antibiotic Resistance. Black bars show the time between the introduction of widely used antibiotics and the development of resistance (data from the Centers for Disease Control and Prevention, 2013).

The rapid spread of antibiotic resistance in microbes infecting both people and livestock presents the specter of future generations once again dying from routine infections we thought we'd conquered. That such a future is creeping back out of the shadows represents a remarkable retreat from our twentieth-century confidence that modern medicine was on the verge of vanquishing infectious diseases.

Humanity won many skirmishes against pathogens in the twentieth century, but today, after decades of indiscriminant antibiotic use, people are once again dying of bacterial infections that were easily treatable a few decades ago. Are MRSA (methicillin-resistant *Staphylococcus aureus*) and antibiotic-resistant *M. tuberculosis* the vanguard of a twenty-first-century bacterial counterattack?

In our zeal to win the war, we've failed to use antibiotics as strategically as we could. In the course of our attempts to kill human pathogens with them, we've also contributed to rearranging our microbiome. For too long we have been taking out our own front line of defense.

The latest realization about the effects of antibiotics is striking indeed. Researchers at Oregon State University report that, in experiments with mice, antibiotics kill more than just bacteria. They also take out cells lining the colon. How is it that antibiotics can kill mammalian cells? Through damaging mitochondria, the miniature powerhouses found in every cell. Recall that long ago mitochondria were free-living bacteria. Apparently, the microbial roots of mitochondria make them vulnerable to certain antibiotics.

Germ theory can't explain the dramatic rise over the past fifty years in chronic and autoimmune diseases that lack an infectious agent. Neither can changes in human genetics—our genes could not change so much for so many of us in just two generations. But what does change that fast is our microbiome. To microbes, a single human generation of thirty years is more than three-quarters of a million generations. By such generational accounting, our species has been around nowhere near that long. Each of our lives represents an evolutionary playing field for microbes that may work in our favor, or not.

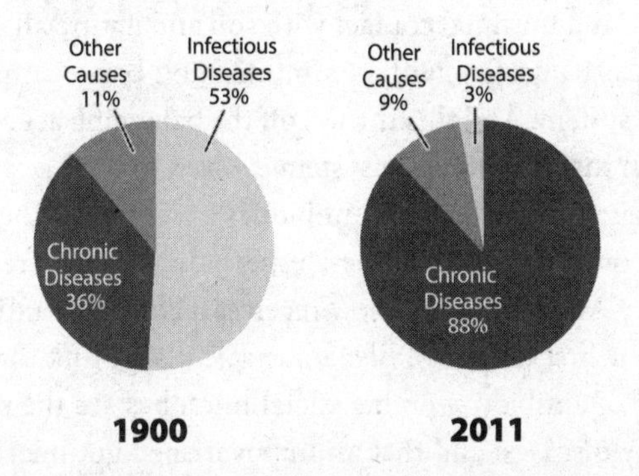

The Big Change. Over the course of the twentieth century, chronic diseases overtook infectious diseases as the leading cause of death in the United States (data from Jones et al., 2012).

Growing up in the 1960s and 1970s, neither of us can recall classmates or friends with severe enough allergies and asthma that it required hypervigilant parents and teachers to help them avoid near-death experiences. We also don't recall today's prevalence of common gut dysfunctions like Crohn's disease and irritable bowel syndrome.

In the past fifty years, researchers have seen not just an uptick in the incidence of gut dysfunctions, but a fortyfold increase, from one in 10,000 people affected to one in 250 people. While our genes may make us more or less susceptible to such ailments, changes in our gut microbiome are increasingly implicated as well.

Gut dysfunctions and autoimmune diseases like asthma and allergies are turning out to be, at least in part, consequences of an immune system gone alarmingly awry. The hallmark symptom of all these diseases is an over-the-top immune response that damages our own cells and tissues.

How does our own immune system turn against us? Increasingly, it seems that a major contributing factor is a severe case of atrophy for our efficient and evolutionarily honed immune system. Without a challenging workout and the help of beneficial microbes, our specialized immune cells and tissue grow lazy, or one might say, hazy. It is the day in, day out saturation of the inside and outside of our bodies with microbes that tones and sharpens the various feedback loops that drive our immune system to learn and recognize microbial friends from foes. A too-clean environment, ultrasanitized food and water, repeated doses of antibiotics, and minimal contact with soil and nature all work against us. These factors interfere with communication between microbes and our immune system. And this throws off the balancing act of meting out inflammation that our immune system evolved to do.

Dousing our inner soil with antibiotics compounds the problem of inadequate contact with microbes, especially when we're young. On average, every American child now receives a course of antibiotics once a year for their first decade of life. Some scientists think that the devastating effects of antibiotics on beneficial microbes are the root cause of inflammatory diseases, and that an impoverished gut microbiome generally spells trouble. With our immune system relying on microbes for information, it's hardly surprising that an off-kilter community of gut microbiota could lead to immune cells misfiring and misjudging who's who. The near-ceaseless stimulation—and modulation—that microbes provide your immune system is at the heart of the way it works for us. As a community, our gut bacteria are vastly more beneficial than harmful—most work with us, most of the time. But when they don't, we sure want antibiotics to work.

Not only did antibiotics save Loki, but they have saved both of us too. Without these miracle drugs, many people over the last fifty-plus years would have suffered unnecessarily and died prematurely. And this underscores the need to question the wisdom of the indiscriminant use of these important drugs given the looming prospect of widespread antibiotic resistance.

What other options do we have? We could try working with the microbial allies that aid our immune system. And a surprising number of novel therapies are doing just that—with remarkable results.

11

PERSONAL ALCHEMISTS

People throughout the ages have sought the key to good health and long life. Along the way, we've prayed to gods, sat in hot springs, and imbibed concoctions. Nobody thought the key to understanding the human body's defenses could be found in a starfish, least of all Russian zoologist Ilya Ilyich (Élie) Metchnikoff.

In the fall of 1882, he and his family moved to Messina, Sicily. Upon arrival, Metchnikoff set up a laboratory in their drawing room. A few months later, in December, he declined an invitation to visit the circus with his family, electing instead to stay at home and continue his experiments. He went out to the garden and pulled a thorn off a rosebush. Then he returned to his laboratory, pulled a chair up to his microscope, and plunged the thorn into the transparent body of a larval starfish. The next morning he watched an amazing scene play out on the stage of his microscope—amoeba-like cells were swarming the thorn. Metchnikoff wondered what he had stumbled upon.

He had previously watched these wandering cells attack and digest particles of red dye he had injected into starfish larva, and now he realized that if these odd, wandering cells could attack something as imposing as a thorn, might they not attack microbial invaders as well? Was this Darwinism operating at a microscopic level? Were the tiniest of struggles at the root of why some animals could fend off diseases better than others? Armed with keen intuition, but no direct evidence, Metchnikoff leapt to the conclusion that such cells were the foot soldiers of the body's defenses.

This was the break that Metchnikoff had hoped for. The thirty-seven-year-old had spent years toiling in obscurity, studying starfish and sponges and enviously following the revolutionary discoveries of Louis Pasteur and Robert Koch. Metchnikoff called the hungry cells he had observed phagocytes. Astonished to find microscopic warfare playing out inside animals, he devoted his life to developing—and defending—the phagocyte theory of immunity. His chance discovery turned the unknown zoologist into a controversial pathologist, culminating in a Nobel Prize decades later in 1908.

Before Metchnikoff's discovery, bacteria were thought to cause inflammation. His observations turned this notion on its head—phagocytes caused inflammation as they fought bacteria. In his view, inflammation was critical to maintaining the health of organisms. And today we know he was right.

Convincing skeptical colleagues that phagocytes ate pathogenic bacteria took Metchnikoff a decade. Along the way, his many clever experiments produced insights about bacteria that formed the basis for another controversial idea.

In the years leading up to and after his Nobel, Metchnikoff became obsessed with the notion that altering the microbial inhabitants of the colon could increase longevity. At the time, most scientists viewed the colon as a putrid cesspool of decomposition, a vestigial organ where microbial inhabitants produced toxins that seeped into and aged the body. Metchnikoff held this view too, initially. He saw the colon as little more than an onboard garbage can that saved our ancestors the trouble of making frequent stops to empty their bowels. Doing so would not have only been inconvenient, but perilous in a world full of predators and enemies. Metchnikoff reasoned that, like all garbage cans, the colon's contents could overflow at times, wreaking havoc on the body until phagocytes arrived to clean up the mess. But why not try and prevent the mess in the first place? What if there was a way to change the microbial population at the source—inside the colon itself—so that noxious compounds and pathogens never accumulated?

Experiments had already proven that the bacteria that made sour milk products like yogurt and kefir produced lactic acid as a byproduct. And lactic acid, it turned out, inhibited other bacteria that contaminated such foods. In the presence of lactic acid, the putrefiers grew slowly, if at all. Metchnikoff suspected that lactic acid producers inside the colon would reduce bacterial toxins and pathogens, and thereby slow cellular aging.

Setting out to find the bacterium that produced the most lactic acid, he turned up *Bacillus bulgaricus*. Metchnikoff subsequently learned that an unusually large number of centenarians lived in Bulgaria, where *B. bulgaricus*–laden yogurt and kefir were staples. The longevity of Bulgarians convinced him that replacing the toxin-generating bacteria in the colon with more friendly bacteria would prolong life.

Though steeped in germ theory, Metchnikoff changed his view of microbes. Microbes were not all bad; some could be good for us. The colon graduated from a sewer to a palace in his estimation. This turnaround presaged what we know today as probiotic therapy. And he walked his talk, drinking kefir every day until he died at the ripe old age of seventy-one in 1916 (at a time when life expectancy in Europe was only about forty years).

While Metchnikoff was one of the first to document the health benefits of probiotics, medical interest was short-lived. Not long after Metchnikoff died, studies suggested that *Bacillus bulgaricus* could not survive the trip through the stomach. By the 1920s and 1930s, probiotic therapies moved on to use another and heartier bacterium found in cultured milk products, *Lactobacillus acidophilus*, with the decidedly more modest goal of treating gastrointestinal disorders, not extending human lifespans.

Soon the dawn of the antibiotic era quickly eclipsed probiotics as the way to safeguard health. Serious research on *Lactobacillus* as a therapeutic agent abruptly ended. Why eat germs for medicine when you could pop a miracle drug that killed pathogens? For decades, little more was heard about probiotics in the Western world. But ideas are shifting once again as we learn about the effects of antibiotics and diet on our internal microbial ecology. Combine a change in the scientific climate with a clever scientist facing a personal health challenge, and a breakthrough might just pop up.

POISON FROM WITHIN

That's exactly what happened when Chinese microbiologist Liping Zhao came across an intriguing paper in 2004 about how the composition of gut microbes in mice might influence obesity. Zhao was carrying a lot of extra weight at the time and wondered if he could somehow change the composition of his own gut microbiota.

Born in the early 1960s, right before the Cultural Revolution, Zhao grew up in a very different China than the one that exists today. Most Chinese lived in rural farming communities, where being overweight, let alone obese, was uncommon. But some forty years later, after college, graduate school, and a two-year postdoctoral stint at Cornell University, Zhao had packed on the pounds.

Worried about his own health and intrigued with the idea that microbes were associated with obesity, Zhao decided to see if changing his diet would change his gut microbiota and help him shed unwanted weight. The idea wasn't such a leap. Zhao had spent his early career in plant pathology, investigating how bacteria could be used to control plant diseases. And he'd grown up eating more of a traditional Chinese diet—rice with lots of vegetables and a small amount of meat. Back then, the so-called Western diet, with its focus on lots of meat, processed foods, and added sugar, fat, and salt hadn't yet infiltrated China.

Zhao returned to eating a traditional diet, focusing on foods long considered medicinal, especially whole grains, Chinese yam, and a squash called bitter melon. These foods, he believed, could positively influence the makeup of his gut microbiome. Zhao's new way of eating paid off. Over the course of two years, the extra pounds fell away—nearly forty-five in all. He dubbed his new diet WTP—W for whole grains, T for the traditional foods, and P for prebiotics (certain foods you'll hear more about later).

He analyzed his stool samples to monitor changes in his gut microbiota and found his new diet especially agreed with one bacterium, *Faecalibacterium prausnitzii*. Initially, this bacterium hadn't shown up in his stool analyses. But at the end of two years *F. prausnitzii* accounted for a whopping 15 percent of Zhao's gut microbiota. This bacterium is of particular interest to those with Crohn's disease or ulcerative colitis, conditions in which the colon is chronically inflamed. When *F. prausnitzii* is introduced, inflammation diminishes.

Hearing of Zhao's self-experiment, a man in need of serious help showed up at his lab, just as young Joseph Meister appeared on Pasteur's doorstep a century earlier. The twenty-six-year-old man weighed 385 pounds. Health conditions associated with obesity plagued him, including high blood pressure, high blood sugar, and high triglycerides, the term for the amount of fat circulating in the blood. They worked out an arrangement. If the man followed the WTP diet, Zhao would measure and monitor the results.

Zhao discovered this man also had high levels of a molecule called lipopolysaccharide in his blood. Although this molecule is found in the cell walls of bacteria that are normal inhabitants of the human gut, it's bad to find a lot of it circulating in the blood. The other name for lipopolysaccharide is endotoxin (poison from within). Endotoxin can also arise from a bacterial infection, and if enough endotoxin gets into the bloodstream, it triggers septic shock (sepsis), the condition that almost killed Loki.

Zhao wasn't too surprised to find that this man had high levels of endotoxin in his blood. Obese people can have endotoxin levels two to three times higher than those who are not obese. But even with no overt symptoms, there is a downside to high levels of endotoxin circulating in the blood. It leads to low-level inflammation throughout the body.

Endotoxin escapes from the gut in several ways. One route out is quite simple—leakage. Even the smallest of space between cells lining the colon is enough for endotoxin (and other colonic contents) to slip out of the colon wall into the bloodstream. This scenario creates "leaky gut syndrome," just the condition Metchnikoff feared, though he didn't know it by name.

The other thing Zhao found in high levels was the source of the endotoxin—bacteria in the *Enterobacter* genus. They composed about one-third of the man's gut microbiome. Not all lipopolysaccharide molecules in the cell walls of bacteria are the same and, although the chemical variation may be minor, the effect can be major. The lipopolysaccharide found in *Enterobacter* species makes such bacteria up to a thousand times more virulent than other endotoxin-producing gut microbiota.

Zhao believed he had found a causative agent among the constellation of factors that lead to obesity. His idea went like this—a Western-style diet promotes high populations of endotoxin-producing bacteria. Endotoxin leaks out of the digestive tract and travels to other parts of the body in the blood. This attracts the attention of immune cells, which sets off systemic inflammation. And lastly, this excessive inflammation leads to metabolic changes that lay the groundwork for obesity. The only problem: Zhao lacked proof.

So, he dusted off Koch's postulates, the step-by-step process to prove that a specific pathogen causes a specific disease. To test his hunch he'd have to isolate the troublemaking *Enterobacter* and see if it was linked to a specific health condition—obesity—as a stand-in for a pathogen-caused disease. He analyzed more fecal samples from the obese man and identified the endotoxin generator as *Enterobacter cloacae*. Once he isolated the organism, satisfying Koch's second postulate, he had to see if introducing *E. cloacae* into another mammal would result in obesity. The mammal of choice? Once again, germ-free mice like those Mazmanian's lab used to pinpoint the effects of *B. fragilis*. But before we find out if Zhao suceeded, let's take a closer look at the interplay between fats and simple carbohydrates (sugars) in our bodies.

THE DOUBLE ROLE OF FAT

While it is easy to confuse dietary fat with body fat, there is more to fat than meets the eye. Not only are there good types of fat we should be eating, but hardly anyone thinks about the basic benefits of body fat anymore. Our fat cells are like a supply depot—a place to temporarily deposit energy and quickly access it later when needed.

At one time body fat was useful in the face of seasonal crop failures, or shortages of wild game and plant foods. For such times a reserve of body fat was our plan B for survival. But few in the developed world have ever experienced severe food shortages, and so we eat our way through the sea of deliciousness that surrounds us, packing our ancestral plan B around our middles and on our behinds. Interestingly, dietary fat does not necessarily lead to more body fat. Excess glucose, or sugar, however, does. Eat too many simple carbohydrates and they'll get converted to fats and added to our plan B stockpile.

Why are we programmed to turn simple carbohydrates into fats? First, our bodies strive to keep the amount of glucose at a steady, modest level in the blood. This serves two purposes—it prevents organ damage and provides us with a reliable energy supply. And fat is an efficient form in which to store and access excess calories. A gram of fat contains nine calories of usable energy, whereas a gram of carbohydrate (or protein) contains only four calories. We can store more energy per unit of weight if it's stored as fat. And when we need to tap our energy depot, fat is converted back into carbohydrates, specifically glucose, the fuel for most of our cells. This is the beauty of body fat—a portable, accessible source of energy. What more could we ask for?

A lot actually. Like a plan B that doesn't mess with our health.

The cells that make up fat tissue are as specialized as those of the liver or the heart. And metabolically speaking, fat tissue is very active, playing a role in everything from regulating blood sugar and hormones to immunity. Recall that cytokines are signaling molecules that help immune cells communicate with one another. Fat cells produce cytokines too. Some interact with the brain, compelling us to start eating if we have hunger pangs. Others help regulate blood pressure, prompt the release of insulin, and urge the liver to release, or hang on to, its stash of glucose. These cytokines act so much like hormones that fat tissue is likened to a second hormonal system.

Interestingly, fat tissue also happens to contain immune cells. Quite a lot of them, as it turns out. Up to 50 percent of fat tissue in obese people is made of macrophages. In nonobese people, macrophages compose only about 5 percent of fat tissue. And compared to lean people, obese people also have a higher number of pro-inflammatory T cells in their fat tissue relative to anti-inflammatory T cells.

When endotoxin from the gut floods into fat tissue, the local macrophages and T cells read it as antigen. And the combination of lots of antigen and lots of immune cells unleashes a boatload of pro-inflammatory cytokines, one of which is interleukin 6 (IL-6). And this brings us back to Zhao, who monitored this cytokine in himself and the obese man. Initially high, their levels of IL-6 dropped over the course of eating the WTP diet. Zhao's research showed but one way that a modern lifestyle undermines our once-brilliant plan B, transforming it from an asset to a serious liability.

PROOF IN HAND

Let's find out if Zhao fulfilled Koch's third and fourth postulates—determining whether or not *E. cloacae* could spur obesity in mice as well as in people. He divided germ-free mice into three groups. The first group was fed high-fat mouse chow, but not inoculated with *E. cloacae*. These mice did not become obese. The second group of mice, also on a high-fat mouse chow diet, were inoculated with *E. cloacae* obtained from the obese man. By the end of the week, the mice in this group started putting on weight, and before long they were obese. Zhao fed the third group of germ-free mice with normal mouse chow, and also inoculated them with *E. cloacae* from the obese man. This group of mice, however, did not become obese.

Zhao then compared the endotoxin levels in the two groups of mice inoculated with *E. cloacae*. Those that ate the high-fat chow had markedly higher levels of endotoxin than the group that ate normal mouse chow. Zhao believed these results fulfilled the third postulate—that introduction of the causative bacterium led to obesity in mice on a high-fat diet. Zhao didn't fully follow Koch's fourth and last postulate, retrieving the causative bacterium from the obese mice. This wasn't really necessary. Unlike Koch, who did not have germ-free mice, Zhao knew that the only microbe in his germ-free test subjects was *E. cloacae*, the one he added. In jumping this last hurdle, Zhao concluded that obesity stems from a combination of two factors, a high-fat diet *and* endotoxin circulating in the blood that bacteria in the gut produce.

Zhao's conclusions are consistent with prior and subsequent experiments on mice, which also show that a high-fat diet leads to high levels of endotoxin, in some cases two to three times higher than in normal mice. Mice are not people, however, and skeptics rightly question the veracity of using rodents to unravel the mysteries of the human body, especially in diet studies.[1]

Still, the results for the obese man who first came knocking on Zhao's laboratory door are striking. On Zhao's WTP diet, he shed 113 pounds over twenty-three weeks, an average of more than half a pound a day. As impressive as those results are, Zhao knew that one human subject was not convincing. So he enlarged his study: ninety-three obese people would follow the WTP diet and Zhao would study the results.

Among the whole grains the study subjects ate were adlay (*Coix lachrymal-jobi*), buckwheat, and oats. The traditional Chinese medicinal foods again included bitter melon and the prebiotics included pectin and oligosaccharides (sources of dietary fiber). After nine weeks, the ninety-three participants changed in much the same way as Zhao and the obese man had—they lost weight, their blood pressure and triglyceride levels (fats) decreased, and their blood sugar declined to a healthy level.

Zhao didn't let the participants off easy. After the initial nine-week phase of the experiment ended, he followed them for another fourteen weeks. All the participants were instructed in how to make meals true to a WTP diet. Zhao continued supplying the bitter melon and prebiotics. At the end of the fourteen-week phase, some of the people once again had elevated levels of endotoxin and had regained some of their weight, suggesting that they had strayed from the WTP diet. Even so, at the end of twenty-three weeks, all the subjects had improved metabolic markers compared to when they had started the trial. In particular, the way their cells responded to insulin was greatly improved, reducing their risk for type 2 diabetes.

Zhao also tracked changes in several biomarkers that are indicative of inflammation (lipopolysaccharide-binding protein, C-reactive protein, and the pro-inflammatory cytokine IL-6). All of these markers were significantly lower at the end of both the nine-week phase and the fourteen-week follow-up phase. Systemic inflammation had dropped significantly among trial participants.

The dietary switch not only changed what people looked like on the outside, but also on the inside. They experienced a decrease in the abundance of gut bacteria from two families that make potent endotoxin (Desulfovibrionaceae and Enterobacteriaceae). And the relative abundance of bacteria increased in another family known to help counteract leaky gut syndrome (Bifidobacteriaceae). Zhao concluded that a change in diet could end the reign of endotoxin-producing bacteria, and thus their contribution to obesity.

While Zhao found that the WTP diet had a positive effect on insulin resistance, Dutch researchers took a more direct route. They transplanted the gut microbiota of lean people into obese people to see if it would affect the insulin resistance of the obese participants. At the end of six weeks, the obese people had a much-improved response to insulin and their gut microbiome still contained many of the bacterial species from the lean donors. But the obese people did not change their diet, and after three months their gut microbiota reverted to the pretreatment composition.

Researchers at Washington University in St. Louis carried out an experiment similar to that of their Dutch colleagues, with a significant twist. To rule out genetic differences, they recruited four sets of identical twins in which one twin was obese and the other lean. Stool samples obtained from each twin were then transplanted into germ-free mice. The mice that received the lean twin's gut microbiota stayed lean and the mice that received the obese twin's microbiota became obese.

This was an important finding. Transplanting gut microbiota is not only possible, but can have profound effects. Even more astonishing was that the researchers also found that the microbiota associated with leanness could take over and displace the microbiota associated with obesity. Knowing that mice are coprophagous, which is to say that they eat one another's feces, the researchers cohoused lean and obese mice. Soon the obese mice became lean, leading the researchers to conclude that the obese mice had been colonized by the microbiota from the lean mice.[2]

These findings show that gut microbiota play an underappreciated role in obesity. It's not only how much we eat, but also what we eat and what lives within us that matters. To more fully understand why this is so, as well as the therapeutic effects of a WTP-type diet, it's helpful to understand how food travels through the digestive tract and gets broken down into different things and absorbed. These factors turn out to have a critical effect on a person's health—and the role their microbiome plays in it.

DIGESTIVE CASCADE

The "location, location, location" maxim of real estate agents applies to the digestive tract as well. Whether Zhao knew it or just guessed correctly, the WTP diet is ideal for delivering the right type of food to the colon. Even Metchnikoff would have been surprised at the importance of what gets into—and what goes on inside—the colon.

To understand the connections between diet, the colon, and one's overall health, it helps to follow the metabolic fate of a meal. But, first, a word about terms. We'll be referring to the digestive tract as the stomach, small intestine, and colon. While the colon is indeed called the "large intestine," this is a misnomer of sorts. It is no more a large version of the small intestine than a snake is a large earthworm. In fact, these parts of the digestive tract do very different things. The stomach might better be called a dissolver, the small intestine an absorber, and the colon a transformer. These distinct functions help explain why microbial communities of the stomach, small intestine, and colon are as different from one another as a river and a forest. Just as physical conditions like temperature, moisture, and sun strongly influence the plant and animal communities that one sees on a hike from a mountain peak to the valley below, the same holds true along the length of the digestive tract.

Imagine you are at a Fourth of July barbeque. You saunter over to the grill, spear several pork ribs, and nestle them next to a heap of homemade sauerkraut. You grab a handful of corn chips and a few pieces of celery. The vegetable skewers look good too, so you add one to the pile on your plate. And what would the Fourth of July be without macaroni salad and pie?

You lift a rib to your mouth and start gnawing. A forkful of sauerkraut mingles well with the meat and you crunch your way through another mouthful. The macaroni squishes between your teeth, but the celery takes some chewing. It all slips down the hatch and lands in the acid vat of your stomach.

Gastric acids start dissolving the bits of food. On the pH scale, where 7 is neutral and lower values are more acidic, the stomach is impressive. Its acidity ranges from 1 to 3. Lemon juice and white vinegar are about a 2. Acidity makes the stomach a pretty inhospitable place for bacteria. For a dark, wet, warm place it's unbelievably sterile. It is an extreme environment meant to dissolve food and anything else that enters it. As far as we know, only one bacterium (*Helicobacter pylori*) thrives in the caustic environment of the stomach.

After the stomach acids work over the ribs, kraut, chips, vegetables, macaroni, and pie, the resultant slurry drops into the top of the small intestine. Right away bile from the liver shoots in and starts working over the fats, breaking them down. Pancreatic juices also squirt into the small intestine to join the digestive party. Your Fourth of July feast is well on its way to full deconstruction, breaking down into basic types of molecules—simple and complex carbohydrates (sugars), fats, and proteins. In general, there is an inverse relationship between the size and complexity of these molecules and their fate in the digestive tract. Smaller molecules, primarily the simple sugars that compose the refined carbohydrates in the macaroni, piecrust, and chips are absorbed relatively quickly. Larger or more complex molecules take longer to break down and are absorbed in the lower reaches of the small intestine.

The sausage-like loops of the small intestine provide an entirely different type of habitat for your microbiota than the stomach. Acidity drops off rapidly and, in combination with all the nutrients, the abundance of bacteria shoots up, to 10,000 times more than that in the stomach. But conditions still aren't ideal for bacteria in the small intestine. It's too much like a flooding river. And understandably so, considering that about seven quarts of bodily fluids, consisting of saliva, gastric and pancreatic juices, bile, and intestinal mucus flow through it every day. And that's not including the two additional quarts of whatever other liquids you consume. The rushing swirl of fluids entrains food molecules and bacteria and carries them rapidly downstream. The constant motion means that nothing stays put for long, so bacteria can't really settle in and contribute much to digestion.

By the middle to lower reaches of your small intestine, the fats, proteins, and some of the carbohydrates in the Fourth of July slurry are sufficiently broken down for absorption and pass into the bloodstream through the intestinal wall. Notice we said *some* of the carbohydrates. A good amount of them aren't broken down at all. Complex carbohydrates have a completely different fate than simple carbohydrates.

Most of the complex carbohydrates in the vegetable skewer, celery, sauerkraut, and pie fruit sailed right through the acid vat of the stomach, and even past the slew of digestive enzymes in the upper small intestine, to land in your colon. Most of the complex carbohydrates in fruits and vegetables are indigestible, at least by you—and every other person on the planet. Your doctor calls them fiber.

In the botanical world, complex carbohydrates are called polysaccharides. These molecules act like rebar, allowing plants to soar skyward like a high-rise building. One such polysaccharide, cellulose, is found in the cell walls of nearly every plant on Earth. Cellulose lends supple strength to the stems of wheat and the trunks of trees. This is why plants can sway in the breeze and survive ferocious winds. By virtue of the great number of plants on the planet, cellulose wins the prize for most abundant biochemical compound on Earth. In the soil, breaking cellulose down into reusable molecules keeps legions of bacterial and fungal decomposers constructively, and perpetually, occupied.

And the same holds true for cud-chewing ruminants. They have a special part of their digestive tract dedicated to housing the microbes that ferment plant polysaccharides. Sometimes it comes before the stomach, as with the rumen of a cow, goat, or giraffe. And other times it comes after the stomach, as with the so-called hindgut of termites, horses, and gorillas. As a fermentation chamber, the human colon pales next to a rumen or hindgut. But it's perfect for breaking down the complex carbohydrates in our omnivorous diet.

True to their name, simple carbohydrates are just a few sugars linked together. Add one sugar to another, like links in a very short chain, and you're done. This makes deconstruction simple and quick, yielding a lot of glucose in a short time period. In contrast, complex carbohydrates consist of hundreds to thousands of sugar molecules strung together. But making a complex carbohydrate doesn't end there. You need to add more molecules as side branches on the main chain. They could be more sugar molecules, amino acids (the precursor molecules of proteins), fats, or some combination thereof. You get the idea. It takes time for digestive enzymes, at least the ones you can make, to find the right location on a polysaccharide molecule and begin breaking it down into simple sugars. Complexity equals time and, in this case, time is on your side.

Let's catch up with the complex carbohydrates that evaded digestion initially. When the slurry from the small intestine drops into the colon, the environment becomes more of a slough and less like a river. The complex carbohydrates and other undigested food molecules settle in the colon and create tranquil pastures for bacteria to graze. With a neutral pH of about 7, the colon is a paradise for bacteria compared to the acid vat of the stomach or the churning rapids of the small intestine, where the pH runs in the range of 4 to 5.

The colon may be the end point of our digestive tract, but it's the beginning for bacteria loaded with the polysaccharide-busting enzymes we lack. Deep within our inner sanctum, microbial alchemists use our colon as a transformative cauldron in which to ferment the complex carbohydrates we can't digest. Inside or outside a body, fermentation is just another way of breaking down organic matter. But it takes the right microbes. For example, *Bacteroides thetaiotaomicron* makes over 260 enzymes that break apart complex carbohydrates. In contrast, the human genome codes for a paltry number. We can only make about twenty enzymes to break down complex carbohydrates.

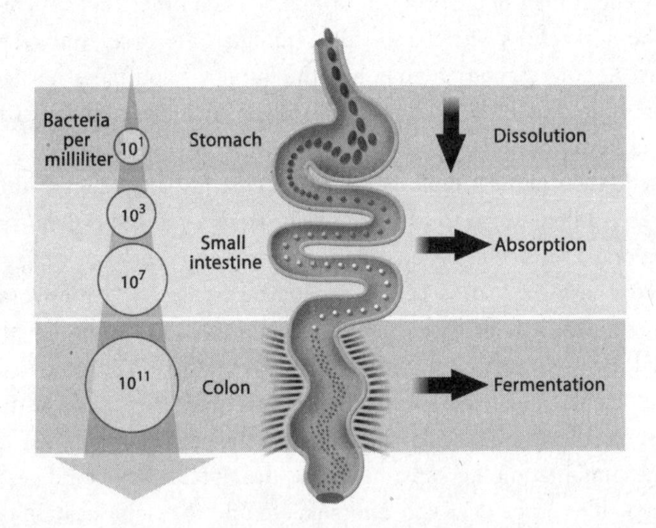

The Food Cascade. Once broken down in the stomach, simple carbohydrates and most fats and proteins are absorbed in the small intestine. Bacteria in the colon ferment complex carbohydrates.

GARBAGE TO GOLD

The colon is a far cry from a lowly garbage can dedicated to collecting and storing what we can't digest. Instead, in this little-loved place one can find a collection of remarkable compounds courtesy of bacterial fermenters in the two phyla that dominate the human gut microbiota—Bacteroidetes and Firmicutes. Their metabolic products, called short-chain fatty acids, or SCFAs, give us a pharmacological cornucopia. Think of short-chain fatty acids as the ultimate in recycling—bacteria thrive on what we can't digest, and then we in turn thrive on their wastes.

From studies on both animals and people, three SCFAs in particular—butyrate, acetate, and propionate—deliver medicinal effects. SCFAs are integral to many processes critical to our metabolism and immune response. They do so by binding to cell receptors found on both immune cells and cells lining the colon. Although researchers don't have a full understanding of all the cellular-level mechanisms by which SCFAs affect health, a bigger picture is crystal-clear. When we fill our cauldron with fermentable carbohydrates, basically mulch, our microbial alchemists turn it into nutritional gold. And as it turns out, SCFAs are also the natural antidote to leaky gut syndrome. They spur the cells lining the colon to pull closer together, in the same way that braces pull teeth closer together. This prevents endotoxin from entering the bloodstream and causing bodywide inflammation.

In one case, Japanese researchers discovered that *Bifidobacterium* species introduced to mice produced acetate, which in turn, greatly reduced the permeability of the gut lining. The improvement was enough to prevent a poisonous compound *E. coli* produces (Shiga toxin) from leaking out of the gut and killing the mice.

The fate of the three primary SCFAs varies. Most of the butyrate stays put in the colon. Well-nourished cells are the foundation for healthy, well-functioning tissues and organs, and the colon is no exception. Cells lining the colon have a high demand for energy and gobble up the butyrate, which supplies 70 to 90 percent of their nutritional energy. Directly absorbing a nutrient like this is highly unusual. Most cells rely on the blood to deliver what they need. Butyrate also induces colon cells to release mucus and antimicrobial compounds central to maintaining a healthy colon wall. And butyrate binds to a particular receptor on colon cells that plays a key role in undercutting and inhibiting cellular processes that lead to colon cancer.

As for the acetate and propionate, they diffuse into the bloodstream and travel to other places in the body—including the liver, kidneys, muscle, and brain. Like butyrate, they provide an energy source for the cells composing these tissues. Propionate, in particular, has another interesting effect on people. It leads to eating less. When propionate docks onto cell membrane receptors in fat cells, they release a hormone called leptin. And when leptin hits the brain, it sends the simple message "You're full. Stop eating."

Overall, SCFAs help optimize and regulate many processes involved with the way we metabolize and use sugars and fats. Low levels of SCFAs due to eating too few fermentable carbohydrates or diminished gut populations of SCFA-making bacteria can lead to a variety of problems, among them weight gain and the development of insulin resistance. Researchers report that the types of metabolic dysfunctions associated with obesity and type 2 diabetes diminish greatly or cease altogether through increasing and maintaining SCFAs at higher levels.

Short-chain fatty acids also appear to play a role in immunity. Tregs are such an important part of immunity that researchers thought there had to be other pathways leading to their development in addition to the polysaccharide A of *B. fragilis*. And given the potential for SCFA production in the colon, they investigated whether there was a connection between SCFAs and Treg development in the immune tissue around the gut.

Through a series of experiments with germ-free mice, the researchers added a mix of SCFAs to the mice's drinking water and measured the levels of Tregs in the gut-associated immune tissue. There were far more Tregs in the mice who received SCFAs than in those that did not. Propionate in particular was linked to the production of Tregs.

In a different study on mice, a research team with Georgia Regents University drilled down to discover how butyrate interacts with immune cells. When butyrate is present in the colon, it binds to dendritic cells and macrophages in the immune tissue surrounding the gut. The dendritic cells and macrophages in turn prompt Tregs to develop. The butyrate-activated dendritic cells and macrophages also spur other immune cells to release anti-inflammatory cytokines.

Again, these are studies on mice, but they suggest a parallel role for bacterial metabolites in human immunity. Consider, for example, that butyrate enemas are a type of therapy for people who suffer from chronic colon inflammation due to Crohn's disease or colitis. The mechanism may be similar to that in mice—butyrate activating the genes that produce inflammation-quelling Tregs. Researchers are actively pursuing development of additional therapies based on butyrate-generating bacteria.

There is a sequence to microbial alchemy in the colonic ecosystem. Bacterial fermentation of complex carbohydrates is greatest in the upper end of the colon because this is where the contents of the small intestine first land. Bacteria that make butyrate live in the upper colon and those that make acetate and propionate live in the lower part of the colon. The butyrate producers generate carbon dioxide as a byproduct, and when concentrations are sufficient, it provides the raw material for the bacteria that make acetate and propionate. The colonic bacterial community is structured around not only what you eat, but through what others in the community make with what you eat.

An effect of fermentation, particularly in the upper part of the colon, is that SCFA production makes the local environment more acidic. And, this turns out to deter pH-sensitive pathogens, most of which cannot tolerate acidity.

Based on genomic analyses of human feces and from snooping around in people's colons, scientists think they've identified only a tiny slice of the bacteria that produce SCFAs. Many, if not most, of the bacteria remain unknown. And we are even more in the dark about their relationships with each other. But given the known effects of SCFAs and endotoxin, maybe we don't need to positively identify all our potential allies to get their message. It stands to reason that feeding the polysaccharide-fermenting SCFA producers, whoever they are, is as fundamental to good health as breathing clean air and drinking fresh water.

12

TENDING THE GARDEN

So how do we feed and care for our colonic inhabitants? Can we change who they are through diet, and trade out the bad for the good? If so, how long does it take? Researchers from Harvard and Duke decided to find out. They recruited ten volunteers, divided them into two groups, and gave each group a different diet to eat. One group, including a lifelong vegetarian, ate a mostly animal-based diet (meats and cheeses) and the other ate a mostly plant-based diet (fruits, vegetables, legumes, and grains). The researchers used DNA analysis to identify which bacteria were present in each person's feces before, during, and after the experiment.

Within a few days, bacteria and the metabolites they produce from breaking down proteins increased in the fecal samples of people on the animal-based diet. Even the microbiome of the lifelong vegetarian had changed. And in the fecal samples of those on the plant-based diet, bacteria that ferment carbohydrates increased in abundance, as did short-chain fatty acids. This study showed that not only does diet change the microbiome, it can do so remarkably fast.

Prebiotics are another name for polysaccharides that bacteria ferment, and the volunteers assigned to the plant-based diet ate plenty of them. In a sense, prebiotics work like the mulch that gardeners lay on top of their beds. But in our bodies they feed our microbial alchemists, and we reap the benefits—if, that is, we eat plenty of prebiotic foods.

To nutritionists, prebiotics are fiber and they bemoan how little most Americans eat.[1] The recommendation for women is about 25 grams per day and about 38 grams per day for men. But few of us, only about 3 percent, come close to doing so. The rest of us consume just a third to half the recommended amount.[2]

Curiously, the value of prebiotics lies in the indigestibility of dietary fiber. Some polysaccharides, like cellulose, are structural, and abundant in the leafy part of a plant. Other polysaccharides serve as a plant's energy depot, like amylose, which is common in root crops like potatoes and carrots. The skin of apples and pears contains yet another polysaccharide, pectin, while onions and garlic are the source of a common prebiotic called inulin. All these polysaccharides provide gut microbiota with a supply of things to ferment that keeps them alive. There are even sources of fermentable carbohydrates that do not come from plants.[3]

But for most people around the world, plants have always been and will likely remain the main source of prebiotics. Humanity's great cereal grains are the seeds of plants in the Poaceae, or grass, family. They are rich in cellulose and also contain lesser amounts of other fermentable carbohydrates. Eaten in their whole form, they make excellent prebiotics, but if refined, they are transformed into simple sugars and absorbed before reaching the colon.

Adding more prebiotics to your diet can support, or even change, your beneficial gut microbiota. But what do you do when something happens to your microbiome? After all, one of the biggest problems with antibiotics is that they can kill a lot of helpful bacteria along with the harmful ones.

This is where probiotics come in. A probiotic is a living bacterial strain or species that delivers benefits once inside the particular part of the body for which it is intended. While prebiotics feed what you've already got, probiotics can help reintroduce what may be missing.

Cultures in the Middle East and Asia knew about the benefits of eating foods containing living bacteria long before Metchnikoff noticed the longevity of yogurt-eating Bulgarians. It was the Turks who introduced the French to yogurt in the sixteenth century. The namesake of France, Francis I, was cured of severe diarrhea when his ally, the Ottoman ruler Suleiman the Magnificent, sent a physician bearing yogurt to the French court.

Probiotic research spans a range of diseases and health conditions from head to toe, including mood disorders, gut ailments, urogenital infections, liver disease, and certain cancers. A 2012 metaanalysis covering seventy-four studies and eighty-four trials that involved more than 10,000 people showed that probiotics can be effective in helping to prevent and treat gastrointestinal ailments such as irritable bowel syndrome and chronic diarrhea. A person's existing health conditions, the bacterial strain, dosage, and mode of delivery, are among the issues now being targeted as probiotic therapies move forward. Attention to experimental design is helping sort out genuine benefits from unsubstantiated claims.

Probiotics are usually thought of in terms of their potential to help out with gut problems, whether due to the aftermath of antibiotics, a bug picked up while traveling, or some type of chronic inflammation. But when Liping Zhao, the scientist who formulated the WTP diet, conducted more experiments on mice, he found that probiotics can help address conditions at the root of other chronic health conditions too.

Most of the probiotics in use today, or those that are the subject of experiments, come from the families that include the genera *Lactobacillus* and *Bifidobacterium*. Bacteria from these two groups in particular are generally quite low in the microbiomes of obese people and mice. To explore the effects of probiotics, Zhao gathered up mice on a high-fat diet, divided them into three different groups, and fed each one different probiotics. Of these groups, two received different *Lactobacillus* species, and the third received a *Bifidobacterium* species (*B. animalis*). For comparison, two additional groups of mice received no probiotics—one ate a high-fat diet and the other ate normal mouse chow. The results proved illuminating.

First, each probiotic had a different effect. *B. animalis* lowered levels of pro-inflammatory cytokines secreted by fat tissue and was more effective than the lactobacilli at lowering levels of endotoxin in the mice. But the *Lactobacillus* species beat out *B. animalis* on another front. They increased levels of acetate, one of the three beneficial short-chain fatty acids that bacteria produce as a fermentation byproduct.

Another change occurred in all three groups of probiotic-fed mice. Despite all eating the same amount of high-fat chow, their fat cells became smaller and fewer macrophages were present in their fat tissue, indicating less inflammation. The mice also became much less obese and experienced improved blood sugar levels and a marked lessening of fat accumulation in their liver. Lastly, Zhao found substantial numbers of each probiotic species in the feces of mice, proving they could make it through the gauntlet of the gut.

The study previously mentioned, in which volunteers either ate a plant-based diet or an animal-based diet, also found evidence that bacteria make it through the digestive tract. While the investigators hadn't intentionally included probiotics in the diets, several types of lactobacilli increased substantially in the fecal samples of the people in the animal-based diet group. Where did they come from?

From age-old sources, it turns out—bacterial cultures used to make cheeses and cure meats. In addition, two species of fungi found in feces were traced to cheeses in the animal-based diet and vegetables in the plant-based diet. And lastly, a plant virus, Rubus chlorotic mottle, even sailed its way through the digestive tract. It was found in the fecal samples of those on the plant-based diet, most likely picked up from spinach.

And this leads us back to Metchnikoff's epiphany—using food as a vehicle for getting probiotics into the body. Cabbages are a popular fermentable vegetable these days. Set lactobacilli loose on fresh cabbage submerged in water and lots of salt, and it will soon be brimming with life. A few lactobacilli can blossom into many in no time at all, as long as they have something to ferment. Eat your sauerkraut or kimchi and some of the lactobacilli join the others down in your colon—and some may turn up other places too.

The vagina has been a focus of much research on using probiotics to treat and cure infections. Several *Lactobacillus* strains have proven quite effective at ousting pathogens and restoring vaginal health. As in the cauldron of our colon, the fermentation of sugars is the foundation of vaginal health too. Vaginal cells provide the fermentable sugars, as well as other nutrients, on which lactobacilli thrive. Sound familiar? Plants use sugar-rich exudates to attract beneficial microbes to the rhizosphere. And when lactobacilli ferment vaginal sugars, they cast off lactic acid, a metabolite that is as critical to vaginal health as SCFAs are to colonic health.

A disrupted vaginal microbiome is one of the most common reasons that women go to a doctor. Doctors know this condition as bacterial vaginosis, and at any one time, an estimated one in three women have it. Some women have symptoms, but a great many don't. Either way, a number of other health issues stem from bacterial vaginosis for women or their partners—preterm labor, infertility, and an increased susceptibility to contracting sexually transmitted diseases, among them HIV, certain herpes infections, and perhaps even HPV. So, all in all, it is highly desirable to have an abundance of indigenous lactobacilli in the vagina.

However, when a woman with bacterial vaginosis goes to see her doctor today, she receives the same treatment she would have received over fifty years ago—antibiotics. Despite the benefit of antibiotics, after an initial infection clears it often recurs. One study found that the cure rate using two mainline antibiotics after four weeks ranged between 45 percent and 85 percent. And after three months, reinfection occurred in up to 40 percent of the women, and by six months half the women had an infection again.

Not only does bacterial vaginosis have a high recurrence rate in the short run, but other maladies commonly crop up after antibiotic use—vaginal yeast infections and urinary tract infections. Researchers think that the treatments for these secondary infections—antifungals to treat yeast infections and more antibiotics to treat urinary tract infections—never allow the vaginal microbiome to reestablish itself to preinfection conditions. So another bout of vaginosis comes on. And another round of antibiotics is prescribed. Here then is another case of a "solution" that fails to solve the original problem and creates additional ones.

The cycle of women clearing bacterial vaginosis only to acquire subsequent infections misses the mark of successful medical treatment. And so, starting in the 1970s, clinical researchers took another approach. Why not try to restore lactobacilli populations? After all, these bacteria have an effective strategy for keeping their house healthy and clean. Lactobacilli stick to the cells lining the vagina, depriving pathogens of a seat at the sugar table. And lactic acid, like SCFAs in the colon, makes the local environment of the vagina more acidic, another deterrent to pathogens. Lactobacilli also seem to communicate and stimulate a just-right immune response similar to the way bacteria in the gut communicate with gut-associated immune cells. Lastly, lactobacilli make their own equivalent of antibiotics—hydrogen peroxide and several other antimicrobials—which knock back pathogens looking to gain a toehold.

Although not all clinical trials followed gold-standard experimental design, they indicate that probiotic therapy is quite effective for treating bacterial vaginosis.[4] Several studies have reported that, after one month, women with an initial infection who received only probiotics had cure rates of about 90 percent, far greater than the roughly 50 percent cure rate for those who received only antibiotics. Moreover, side effects of probiotics are rare and, if they occur at all, are nowhere near as troublesome as those of antibiotics. Other experiments have combined probiotics and antibiotics and achieved better cure rates than those using antibiotics alone. This is perhaps the best strategy—combine the fast-acting kill power of an antibiotic with rapid restoration of the vaginal microbiome to prevent pathogens from setting up shop all over again.

Given the vulnerability of solely relying on antibiotics, and the opportunity to prevent the slew of other serious health problems associated with bacterial vaginosis, a number of prominent researchers openly lament that probiotics remain so far removed from mainstream gynecological practice.

A rather unusual way to acquire new microbiome players delivers an impressive cure rate—from the bottom up. In 1958, doctors at the Denver Veterans Administration Hospital published the results of the first fecal microbiota transplants (FMT) after they administered enema infusions of feces from healthy donors to four patients suffering from life-threatening diarrhea. All four rapidly returned to health, prompting the doctors to conclude that the method deserved serious clinical evaluation. Few of their peers took them up on the idea, however, and for half a century the curious practice remained a therapy of last resort for patients unresponsive to antibiotics.

Sometimes antibiotics so decimate populations of gut microbiota that bacterial scourges proliferate and aggressively colonize the intestinal tract. *Clostridium difficile*, one of the worst, causes horrendous diarrhea that can turn life-threatening. Over the past several decades, *C. difficile* infections have increased dramatically, with between half a million and 3 million cases occurring annually in U.S. hospitals at a cost of over $3 billion. The increased prevalence of C. *difficile* infections has led to recommendations for doctors to suspect it for patients who develop chronic or severe diarrhea after antibiotic use.

When antibiotics kill off, or substantially decrease, the beneficial bacterial population in the intestinal tract, it's much like creating an artificial clearing in a tropical forest. There is no way that patch of ground will remain empty for long.

The basic idea of an FMT is to use microbial ecology as a medical practice—reseed the bare patch in a patient's gut with beneficial bacteria to prevent C. *difficile* from doing so. The mechanism by which the transplanted microbiota displace C. *difficile* isn't entirely understood. But it works. Recent reviews summarizing the available data from hundreds of patients in dozens of studies report cure rates of around 90 percent for those with C. *difficile* infections unresponsive to antibiotics.

The striking success rate of fecal transplants matches that of common medical practices few would dream of going without, like childhood vaccination for polio. And fecal transplants far exceed the lackluster performance of annual flu shots, which have success rates around 60 percent in the best years.

At the same time, researchers and doctors in the FMT field emphasize the need to thoroughly screen donors. Traditional infectious diseases are a primary concern, but so are other health conditions such as obesity and type 2 diabetes, both of which have been increasingly linked to the composition of the gut microbiota.

The last major (scientific) impediment to widespread adoption of FMTs was removed in 2013, when randomized controlled trials (the gold standard of medical research) provided convincing evidence that the procedure did indeed work. In fact, it worked so well that the trial was closed down after assessing the results on the first 43 of 120 planned patients. The 94 percent cure rate for those receiving the treatment was so far above that for patients receiving more conventional treatments (23 percent to 31 percent) that the medical safety board overseeing the trial terminated it in favor of using FMT as the new standard of care.

Other studies have shown that the gut microbiota of fecal transplant recipients shift substantially and persistently. For most recipients, populations of C. *difficile* dropped significantly, and major increases in

Bacteroides species occurred in patients in which these beneficial bacteria had been notably lacking. Recipients' postprocedure gut microbiota came to parallel donors' intestinal microbiota, showing that the introduced bacteria successfully repopulated the recipient's gut. The phenomenal effectiveness of FMT, even months after treatment, is a testament to the medical power of applied microbial ecology.

The demonstrated ability of FMTs to clear *C. difficile* infections and alter gut microbiota has led to new efforts to explore potential applications of the technique in the treatment of other diseases, like autoimmune disorders, obesity, diabetes, and multiple sclerosis. The approach is already evolving. New delivery methods, like freeze-drying feces and putting them in an oral capsule, will undoubtedly increase adoption rates.[5]

GRAIN WRECK

There is abundant medical evidence that diet greatly influences health, and new microbiome science is helping to show why this is so. In a nutshell, for better and worse, what you eat feeds your microbiome. In thinking about such connections, the seeds of the world's major cereal crops (grains) are a good place to start, as they account for the lion's share of what the world eats. Lucky for us, grains offer a nearly perfect nutritional package. Whether wheat, barley, or rice, all have the basics—proteins, fats, and carbohydrates, along with many of the vitamins and minerals essential for health. They also contain many phytochemicals. So why do grains get such a bad rap these days? Much of the problem lies with what we do to grains after they are harvested.

Just as we now know that some fats are better for us than others, the same is true for carbohydrates. As we have seen, the sugars found in simple carbohydrates are rapidly absorbed in the small intestine, whereas the sugars that make up complex carbohydrates travel on to the transformative cauldron of the colon. This simple difference leads to two major problems with a low-fiber diet: bacteria that produce SCFAs in the colon don't get enough to eat, and too much glucose enters the bloodstream too fast. It's a recipe that fuels inflammation and the onset of type 2 diabetes, obesity, and other maladies. But how is it that the bulk of what humanity now eats could undermine our health?

It has to do with the structure of a plant seed. Consider a grain of wheat. The outer seed coat (the "bran") and the inner embryo (the "germ") are small in terms of the overall seed weight. The bran composes 14 percent of the total weight, while the germ adds another 3 percent. Despite their low weight, these two parts of a seed are packed full of nutrients. In addition, some very tough polysaccharides are found in the bran. The bulk of the seed is called the endosperm. It makes up the remaining 83 percent of a seed's weight. Endosperm contains most of the simple carbohydrates and nearly all the proteins found in a seed (including the two in wheat that become gluten in the bread-baking process). In effect, the endosperm is like the placenta of a plant. Had the seed fallen to the ground and germinated, the carbohydrate-rich endosperm would have provided for the seed until it grew roots and leaves and could feed itself. While a sprouting plant clearly needs this type of supercharged energy supply, it's not so good for us in large amounts.

Nature made the bran indigestible for a reason. Many seeds travel through the gut of a bird or mammal. It's a great strategy for a plant to spread itself around. The tough-as-nails outer coat of bran protects the embryo on its journey through an animal so that a seed arrives at its destination intact and ready to germinate.

When someone says a grain is "refined," it means that the bran and the germ are stripped out when the seed is milled. Only the endosperm remains. Grind up the endosperm of wheat grains and you have white flour, which to your small intestine is an easily absorbable sugar. In addition, refined grains contain more gluten per unit of grain than whole grains. This means that a person who eats mostly refined grains consumes more gluten per unit of grain than someone eating a comparable amount of whole grains.

Part of the reason grains are refined is because the fats go rancid—things made from refined flours last longer. Also, bakers don't like bran in flour because it interferes with the elasticity of dough and inhibits rising. Removing these pesky parts of a grain solves those problems. But it causes a whole host of new ones for our bodies. When a seed goes through milling and processing, its perfect nutritional package falls apart.

All cereal grains are amenable to refining. It's the basis for all those eye-popping choices of boxed and bagged items in grocery stores around the globe, especially in the Western world. Refine corn, add some fats back, toss with salt, and you get the perfect tortilla chip. Do the same with wheat and you can make a fine cracker or bread.

Looking back at carbohydrate consumption over the last century reveals some interesting trends. Americans ate about the same amount of total carbohydrates in 1997 as we did in 1909—just not the same kinds. Over this time period, the proportion of carbohydrates from whole grains dropped from more than half of what we consumed to about a third. From 1909 to about 1945, the American diet was relatively high in both total carbohydrates and fiber. Then a new trend emerged in the years immediately following World War II. Both total carbohydrate consumption and the proportion of carbohydrates consumed as fiber dropped significantly. But starting in the 1960s, total carbohydrate intake began increasing while fiber intake remained flat. By the mid-1980s the rate of overall carbohydrate consumption had risen steeply, back to 1909 levels—only with less fiber content, and far more simple carbohydrates (sugar). The percentage of carbohydrates from corn syrup that people consumed increased dramatically from 2 or 3 percent in the early 1960s to about 20 percent by the mid-1980s. In other words, we now eat far less fiber and a lot more simple sugars than at any time in human history.

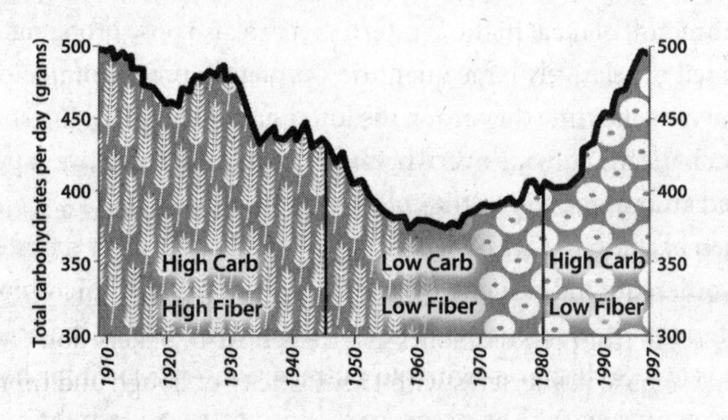

Changing Carbohydrates. Trends in the daily consumption of total carbohydrates and the relative amount consumed as fiber in the United States from 1909 to 1997 (data from Gross et al., 2004).

The difference between what happens when you eat simple carbohydrates and when you eat complex ones helps explain why the people on Zhao's WTP diet, and Zhao himself, experienced many of the changes that they did—like improved blood sugar and less inflammation. The small intestine and colon handle a whole grain very differently than they do a refined grain. When carbohydrates remain bound together with other molecules in whole foods, it takes longer for enzymes to find the carbohydrates and start breaking them down. It's like trying to open a cardboard box triple-wrapped in duct tape versus a box with an easy-open pull tab. Also, the sugar molecules from whole grains have to jockey for space with the protein and fat molecules to make contact with the absorptive cells in the small intestine, further slowing the sugar-absorption process. Plain and simple, when whole grains remain intact, your body absorbs the sugar component at a markedly slower rate.

In contrast, refined grains release a veritable fire hose of glucose, which our small intestine dutifully absorbs and passes on to the bloodstream. This sends insulin charging out of the pancreas to shuttle glucose from the blood into cells. But using cells as a place to endlessly stockpile sugar only leads to other problems, like organ damage. And so, our body solves this problem—by converting excess sugar into fat and moving the overage into depot-like fat cells. When we need the energy, like in the middle of the night long before breakfast, it's there for our use. But an abundance of refined carbohydrates converted into fats overshoots what we need in our plan B energy supply.

The amount of meat in the Western diet can also pose problems. When consumed in relatively large quantities, proteins are not completely broken down by the time they reach the lower end of the small intestine. Eat too much meat and your overwhelmed small intestine delivers partially digested animal protein to the colon.

When bacteria in the colon encounter intact or partially digested protein, a different kind of alchemy gets underway. Colonic microbiota prefer to ferment polysaccharides, but by the lower third of the colon, the supply is low. So bacteria switch to putrefaction—the term for how they break down proteins. The name itself hints at the undesirability of the metabolites they cast off.

The problem with putrefaction stems from some of the elements of which animal proteins are made—a fair bit of nitrogen and small amounts of sulfur. Ammonia, nitrosamines, and hydrogen sulfide probably don't mean much to the average person. But they are among the nitrogen and sulfur-containing compounds that bacterial putrefiers create. These compounds pack a toxic punch to cells lining the colon. They interfere with the uptake of butyrate, which deprives colonic cells of the energy they need to keep the colon functioning in top shape. The spaces between cells begin widening and leaky gut syndrome sets in. Undernourished cells start falling down on the job and wastes accumulate inside the cell, which gums up other cellular operations. Goblet cells slow down mucus production, making the colon lining more vulnerable to pathogens and physical damage. This is not a trivial point. The colon is a busy place and the cells lining it constantly regenerate throughout a person's life. If cells aren't regularly replaced, the effects are somewhat like a house that goes unmaintained. Lots of little problems add up to bigger problems, and eventually the house starts to fall apart.

In particular, the nitrogen-containing compounds that bacteria make from undigested proteins can monkey-wrench cells lining the colon. These compounds can glom onto parts of DNA found in certain genes, altering the genes' activity. So if the gene codes for an enzyme, the enzyme isn't properly made and doesn't do what it is supposed to do. Or, if the monkey-wrenched gene's job is to turn on other genes, it may fail to do so.

The bottom line is that the meat-for-every-meal philosophy of the Western diet can put too much partially digested protein in the wrong place. And for reasons not completely understood, undigested proteins from red meat appear to yield among the most harmful byproducts. Occasional or low-level exposure to nitrogen and sulfur-containing compounds isn't a big deal. But chronically bathing colon cells in the bacterial byproducts of protein putrefaction takes a toll over the years. This may explain why colon cancer comes on late in life and mostly occurs in the lower part of the colon where protein putrefaction occurs.

Other problematic byproducts are made in the colon. Eating lots of fat stimulates the liver to produce bile and deliver it to the small intestine. We need bile. It acts like a detergent and breaks fats into smaller molecules so they can be absorbed. Almost all of the bile used in the small intestine gets transported back to the liver after fats are sufficiently broken down. The key word here is almost. About 5 percent of bile secretions keep moving down the intestinal tract to land in the colon. So, people who eat lots of fat secrete more bile to break down the fats, which means more bile ends up in the colon. But guess who gets ahold of this bile and transforms it?

Our colonic microbiota. They convert bile into decidedly vile compounds called secondary bile acids. And like putrefaction byproducts, secondary bile acids are toxic to cells lining the colon. They damage DNA and cause cells to grow abnormally. And whenever abnormal cells crop up, they create the potential for tumors.

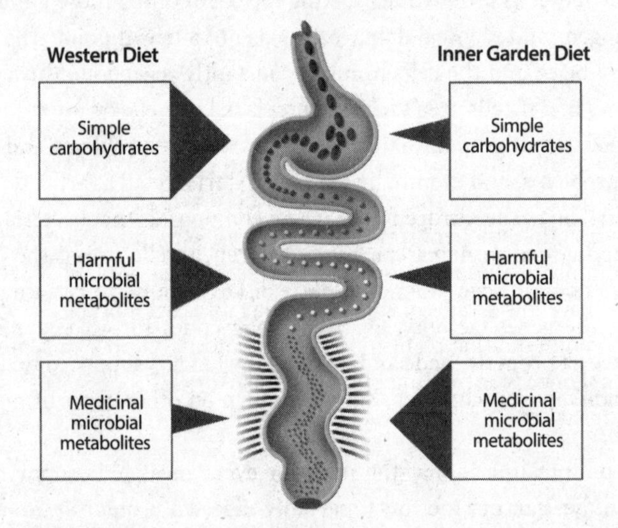

Your Diet Matters: Different diets have different effects on the gut microbiota. Arrow sizes represent the relative amount of dietary nutrients or microbial compounds delivered along the digestive tract. A diet rich in complex carbohydrates yields the highest level of beneficial microbial metabolites.

Curiously, there are indigenous diets, like the protein-rich diet of the Inuit and the fat-rich diet of the people of the Mediterranean island of Crete, that seem like they should be unhealthy. The resolution to this paradox lies in other aspects of these diets. The prodigious amounts of cold-water fish and caribou that the Inuit eat are sources of anti-inflammatory omega-3 fatty acids. So although the Inuit diet is indeed short on plant foods, their primary diet is rich in anti-inflammatory compounds that take the place of bacterially produced anti-inflammatory SCFAs. Likewise for the people of Crete, who eat an extraordinary amount of fat in the form of olive oil—almost thirty liters per person annually.[6] So, what might counter the secondary bile acids needed to digest so much fat? Most likely it's the rich stew of phytochemicals found in olives, along with wild greens called horta that Cretans consume in large quantities. Add more plant foods to this olive oil– and horta–rich diet, and it's likely that the gut microbiota of Cretans churn out impressive quantities of SCFAs that counter the harmful effects of secondary bile acids.

THE OMNIVORE WITHIN

As adherents of the paleo diet like to remind us, humans have long eaten meat. They stress that meat is a fabulous source of many nutrients, especially if the animals being eaten were raised without antibiotics and allowed to follow their normal way of eating. Vegetarians and vegans also admonish us, pointing out that people who eat a plant-based diet generally have lower rates of cardiovascular disease and type 2 diabetes. They also point out that plants possess what animals don't—an astounding arsenal of cancer-fighting phytochemicals.

In other words, both of these countervailing dietary perspectives—paleo and plant-based—contain more than a grain of truth. So consider another perspective. Combining elements of each diet makes a lot of sense given what our colonic microbiota do with the meat, fats, and plants we eat.

Here's how it might play out. Imagine the putrefaction byproducts from undigested meat and secondary bile acids soaking the cells lining the colon. DNA mutations occur and a few abnormal cells start regenerating. A tumor begins growing. The tumor cells gain the upper hand, ignoring instructions from killer T cells to self-destruct. But follow this scene with a tsunami of butyrate, and colonic cells perk up and communicate more effectively with the immune system. Additional killer T cells arrive and take out the renegade cells. Prodigious amounts of undigested complex carbohydrates from plant foods enter the colon, dislodge secondary bile acids and mop them up, thereby reducing contact between these carcinogens and the colon lining. Normal cell growth and mucus production resumes. Health is maintained in the cauldron and thereby the body at large.

This scenario is ingenious from both a health and an ecological perspective. The fiber fermenters have solutions for problems the protein putrefiers create. Plus, everyone in the cauldron gets fed—with either complex carbohydrates, or the castoffs of undigested proteins and leftover bile acids. So long as the byproducts of the fiber fermenters prevail, the colon serves as the medicine chest Metchnikoff envisioned rather than the toxic dump he feared.

We are the most omnivorous creatures on the planet, with a vast array of domesticated crops and animals and wild foods at our fingertips. There is hardly anything people don't eat—the blubber of whales, the intestinal lining of pigs, caterpillars, rotten fish, raw fish, and seaweed, to the more mundane items like meat, dairy, bread, fruits, nuts, and vegetables. Yet many diets and diet gurus shun our inner omnivore. Instead, we are constantly urged to eat a narrow (and ever-changing!) slice of omnivory. Ideas for what we should eat have swung like a pendulum—more toward meat, or more toward vegetables, away from fats, then toward certain kinds of fats, toward whole grains, now away from all grains.

No wonder so many of us are either sick or tired, or both. Perhaps it's worth focusing on what to feed our inner omnivores. It would look a lot like the plate Heidi laid out. The mechanics are pretty simple. Pick a modest-sized plate and make meals using vegetables, legumes, leafy greens, beans, fruits, and unmilled whole grains as the main ingredients. Add some meat if you want and dollops of healthy fats on the side or sprinkled through the plant foods. Desserts and sweets are special, so save them for the special times.

Understandably, special dietary considerations apply to people with gut dysfunctions or who are diabetic or allergic to specific foods. But for most of us the key to healthy eating may be as simple as balance and diversity—and sidelining refined carbohydrates.

We realize a diet like this doesn't lend itself to being packaged and sold. It emphasizes how to think about food in the context of one's microbiome, rather than prescribing a narrow choice of foods or simply counting calories. It's far from sexy and certainly not earth-shattering. But it adds up to what may be the best advice about how to eat that anyone will ever give you. Just think about the alchemy going on in your microbial cauldron. You want your fiber fermenters to churn out far more of their nutritional gold than what your protein putrefiers and bile acid modifiers conjure up. To keep the fiber-lovers on top, fill the cauldron *every* day with fermentative fodder so that it bubbles with things that are good for you. In short, as we all ponder what to eat, it would be a good idea to realize who we are really eating for and what they do with what we eat.

When Anne decided to change our diet, I wasn't thrilled. She had an ace in the hole, though. Cancer lent gravity to her argument and got me thinking more seriously about what I ate. I thought I ate well—most of the time anyway. And while I'd scarf down burgers, pizza, and tortilla chips when opportunities arose, I also thought I ate my share of salads and other vegetables. But I now see that my normal diet included way too many simple carbohydrates—lots of bread, crackers, beer, and wine.

Putting in our first vegetable beds set us on the path to change our diet. Initially, we weren't thinking of cancer or cholesterol or anything like that. Awash in homegrown produce, however, we began eating a lot more vegetables as main courses. And because we ate more things from the garden, we ate less meat, cheese, and bread, and hardly any boxed things from the cupboard.

And after a while, by changing what I ate and deciding to walk the two-and-a-half-mile round trip to work each day, my health dramatically improved. Before, I had high blood pressure, high cholesterol, acid reflux, and chronic gut issues. I'd been taking a drug for cholesterol and little purple pills for acid reflux. My doctor had even talked to me about maybe getting on blood pressure medication too. After a year on our new diet, my blood pressure and cholesterol fell within the normal range. My acid reflux disappeared, as did recurrent bouts of diarrhea. I lost a lot of weight too, about twenty-five pounds. No longer on any medications, I remain amazed at how changing my diet to cultivate the microbial garden in my gut greatly improved my health.

13

COURTING ANCIENT FRIENDS

With surgery behind her, and our new dietary plan in full swing, Anne was inspired to plant different kinds of cancer-fighting crucifers, kale in particular. The big stout-stemmed fan-like leaves grew like crazy in Seattle's cool weather. But just how nutritious was this kale? We decided to find out.

It was early July, so we checked the weather forecast and picked the coolest day of the week. Late on the appointed day, we pulled a wide-mouthed thermos out of the freezer, clipped a total of twelve leaves from several red kale plants, and carefully placed them in it. Then we sped off to the nearest FedEx office and arranged for overnight shipping to a food lab in Portland, Oregon.

A few days later we got the results and compared them to the U.S. Department of Agriculture's National Nutrient Database, a reference that lists standard nutrient levels in food. Despite some unavoidable nutrient loss due to heat and shipping, our greens stacked up. The macronutrient concentrations (P and K) were quite similar to the reference values, but our kale had twice as much calcium (Ca) and zinc (Zn), and four times the amount of folic acid (vitamin B9).

Why was our homegrown kale more nutritious than the average store-bought variety? We hadn't added any fertilizers—just heaps of compost, organic matter. Had this made the difference? As it turned out, we weren't the first to wonder. Many of the early pioneers of organic agriculture had the same question, but they lacked explanations for how soil health influenced food quality.

Scientists are a famously skeptical lot. If you can't explain how something works, you won't gain many converts. Most will stick to the conventional wisdom of what they were taught. A good example is the response of geophysicists of the 1920s and 1930s to German meteorologist Alfred Wegener's outrageous idea that continents move around. Geophysicists contemptuously dismissed the notion outright because Wegener could not identify a mechanism to explain continental drift. Wild ideas without plausible mechanisms don't gain much traction in scientific circles.

Decades later, with the help of new technologies, geologists were able to compile compelling data pertinent to Wegener's theory, bringing to light a sensible mechanism—plate tectonics. Once scientists understood the mechanics of how ocean basins split apart and continents collided, it all came together. Decades after Wegener died, his wandering continents explained mysteries like how mountains rise, and why Africa and South America look like neighboring pieces of a jigsaw puzzle. Today, the discoveries unfolding about microbial life and its influence on the health of plants, soil, and ourselves are producing another seismic shift in perspective.

NATURAL PROPHETS

As far back as the 1930s, Sir Albert Howard championed the idea that microbial life promoted not only soil fertility, but human health as well. But he couldn't explain the mechanisms through which this happened, so mainstream scientists considered his views as speculative (at best). Yet the early ideas of Howard, and a few like-minded peers, appear prophetic now that gene sequencing can tell us more about the composition of microbial communities and their activities in the soil and in our bodies.

In Howard's view, a diet of crops and livestock raised on fertile ground provided a full complement of nutrients. Crops and livestock grown with chemical fertilizers on degraded soils inherently lacked critical nutrients because of disrupted associations between plants and their microbial partners. Although he was unsure of the way agricultural practices translated into human health, Howard was confident that they did, and that soil life lay at the root of the connection.

Feeding his suspicions were curious cases like that of a twenty-three-year-old Irishman who arrived in England during the run-up to the Second World War. He was healthy upon arrival but succumbed to jaundice after just two months on an English diet. In his new home, the young man ate mostly meat sandwiches on white bread, and the occasional egg. In Ireland he had eaten fresh foods—potatoes, milk, vegetables, fish, eggs, and occasionally meat. The dramatic change in diet, Howard concluded, abruptly led an otherwise healthy man to disease.

Another telling example of the role diet played in human health came from the experience of a large boys' school near London. The school had both boarders and day students, and the school staff grew many of the vegetables the students ate. Howard followed what happened when the school changed the way its vegetables were grown, switching from chemical fertilizers to Indore-style compost. Soon, the rampant colds, measles, and scarlet fever that plagued the school were dramatically reduced to isolated cases brought in by the day students. Howard concluded from this that fresh foods, grown in fertile soils, promoted human health.

This wasn't an isolated result. A paper published in the June 8, 1940, issue of the journal *Nature* reported on a similar experiment conducted at Mount Albert Grammar School, a boys' school in New Zealand. When this school switched from food grown in chemically fertilized soil to organic foods grown with no fertilizers, the chronic nasal congestion that plagued the boys cleared up, and the incidence of colds and flu dropped noticeably. The *New York Times* noted that the school's students fared far better than others during the 1938 measles epidemic in a three-paragraph description of the study.

Such cases convinced Howard that the key to maintaining human health lay in returning organic matter to the land, in patterning agricultural practices on nature's cycle of decay and renewal. Though certain that soil bacteria and mycorrhizal fungi supplied key nutrients to plants, he didn't suspect that plants made exudates and traded them with microbes to get other nutrients. And, without an identified mechanism, most academic scientists considered him delusional.

Howard did have some supporters though, chief among them influential English agronomist and farmer Lady Eve Balfour. In the years leading up to and following the Second World War, Balfour studied soil life and its effects on crop quality and yields. On her farm and others, she noticed that adding composted animal manures to soil fostered beneficial soil life, and that this, in turn, helped create and maintain fertile soil. She came to the same conclusions as Howard—organic matter and soil life formed the basis for soil fertility at the root of health in plants, animals, and people.

As the reigning queen of organic agriculture in the 1940s, Balfour combined her own observations with those of the medical community and the work of Sir Albert Howard to argue that healthy soil was the thread connecting health in plants, livestock, and people. Vibrant microbial communities in the soil translated into healthy food and, for those who had access to such food, healthy people.

Balfour's seminal 1943 book, *The Living Soil*, made the case that soil be treated and managed as though people's lives depended on it. Pushing her idea beyond the world of soil science into agriculture and public health, Balfour advocated the still-visionary notion of merging England's ministries of agriculture and health to ensure the provision of fresh, nutritious food for the British public. She envisioned soil scientists working alongside physicians in hospitals and clinics.

Balfour and Howard's insights into the fundamental connections between soil health and human health were sidelined in the aftermath of World War II. Industrialists were busy pivoting factory production of tanks to tractors, munitions to fertilizers, and poison gas to pesticides and herbicides. Interest in the role of soil health in soil fertility faded as the widespread availability of affordable agrochemicals and equipment took center stage.

But even as parts of the defense industry metamorphosed into a burgeoning agrochemical industry, scientists voiced concern about the declining nutritional value of food grown using industrial practices. One of the more vocal, University of Missouri agronomist William Albrecht, warned about reliance on foods rich in calories but poor in nutrients. He forecast that under industrialized agriculture soil health would decline, with human health following on its heels.

Albrecht used his position as president of the Soil Science Society of America to share his view that soil is a nation's most important resource. And like Howard, he believed organic matter fueled the microbial populations that filled the niche that no other organisms could, breaking nutrients back down into simpler forms that could be used all over again to nourish new life. Healthy microbial life was essential to ensure that mineral-derived nutrients in soil organic matter made their way back into more crops. Nature's secret for nourishing future generations—soils rich in organic matter and abundant soil life—kept nutrients in biological circulation.

Albrecht considered the steady return of organic matter to the soil as a practical, rather than a radical, idea. Everyone knew that crops grew better in soil rich in organic matter. Corn Belt farmers harvested twice the yield from fields with twice the soil organic matter. Indeed, the organic-matter content of the soil provided a rough, but reliable, index of land value.

And Albrecht pointed to compelling evidence that returning organic matter to fields was the key to sustaining fertility. A study in central Missouri had compared soil organic matter in virgin prairie with nearby corn and wheat fields harvested continuously for sixty years without the addition of organic matter. Though negligible erosion had occurred, more than a third of the soil organic matter was missing from the cropped fields compared to virgin-prairie soil. Likewise, continuous planting had reduced available soil nitrogen by about a third over the thirteen-year study. This and other studies led Albrecht to the alarming conclusion that without somehow retaining, or adding, organic matter to soil, regular tillage would greatly degrade soil fertility.

Few shared Albrecht's concern. Declines in soil organic matter happened slowly, and U.S. fields still produced profitable crops. So why worry? Besides, farmers saw that fertilizers could immediately pump up yields on degraded land. Albrecht thought this made for a perverse incentive—a farmer could get away with losing the soil organic matter that sustained fertility and still reap higher yields thanks to the magic of cheap fertilizers liberally applied. At least for a while.

Nothing if not obstinate, Albrecht continued his crusade, taking Howard's longer view and arguing for rebuilding soil organic matter. Noting how organic matter was five times more effective than clay at releasing nutrients to plants, he advocated returning crop stubble and organic wastes to the fields. Cycling organic matter back into the earth was the key to sustaining high yields of healthy crops.

Just as Howard and Balfour believed, Albrecht thought that bacterial and fungal crop diseases would thrive in malnourished crops growing on depleted soil. He advocated for the still-overlooked (and eminently sensible) suggestion of granting farmers tax credits if they made long-term investments in rebuilding soil organic matter, and thereby soil fertility. He also promoted the view that the health of Americans rested on the health of their soil.

Albrecht's message that fertilizers were no substitute for healthy soil, and that soil-related nutrient deficiencies underlay many human health issues, did not go over well with an agronomic establishment striving to industrialize food production and bolster yields on degraded land. Scientific colleagues and agribusiness interests alike attacked Albrecht for straying beyond the bounds of his soil science training. Worrying about the health of animals and people was the job of veterinarians and doctors. Unable to explain how the connections worked, the academic community largely dismissed the idea that human health was intrinsically linked to soil health.

Albrecht charged headlong into controversy when he tied patterns in the dental records of 70,000 Second World War–era sailors in the U.S. Navy to regional patterns in soil fertility. In those days, most people ate locally grown foods, so it wasn't much of a stretch to compare the condition of sailors' teeth with the fertility of the soil from where they hailed. What Albrecht found was that sailors raised on the fertile soils of open prairies in the Midwest had fewer cavities and missing teeth than those raised on the degraded soils of the Southeast. Albrecht also noted how areas with calcium-deficient soils had higher draftee rejection rates.

He was particularly concerned about soils in which crops were fertilized with the three best-known fertilizers—N, P, and K. As plants grew, they also incorporated naturally occurring mineral elements, such as copper, magnesium, and zinc. Albrecht asserted that, over time, renewing only N, P, and K, but not trace minerals, would lead to less nutritious food. In other words, intensive chemical fertilization could lead to high yields of mineral-poor crops. And whether in plants or people, deficiencies in essential minerals meant malnutrition, as surely as insufficient calories did.

Albrecht's ideas came together in the late 1940s when he called for a national initiative to restore the health and fertility of America's "worn out" soils. His views were not welcome among agronomists hitching up to the profitable agrochemical bandwagon. And it didn't help that Albrecht made a serious mistake. He vocally advocated for a particular calcium-to-magnesium ratio as the universal ideal for plant health and growth in all soils. This attractively simple idea promised that properly adjusting this ratio would ensure thriving crops. Not only did this prove incorrect, but he dug himself in deeper by dismissing the influence of soil pH on plant growth. It turns out that soils are quite variable and pH matters a lot. Though his idea worked in some circumstances, even his followers found that his magic ratio was not the answer for all soil types, crops, and climates. Albrecht's detractors seized on this misstep and used it to attack his broader ideas.

Undeterred, Albrecht continued his crusade, becoming particularly concerned about how soil health influences the amount of proteins and carbohydrates in animal feed and food crops. He distinguished "go foods" rich in carbohydrates, but lacking both complete proteins and a full complement of mineral nutrients, from "grow foods" that had more favorable protein and mineral content. The latter, Albrecht asserted, came from healthy soils. Arguing that routine consumption of go foods would make people overweight, he astutely anticipated the modern obesity epidemic. Albrecht also predicted that it would take half a century to work out the connections between soil health and human health.

A LITTLE MEANS A LOT

Long neglected, the beliefs of these early prophets of organic agriculture are proving remarkably prescient as scientists today unravel the myriad ways in which communities of microbes influence the cycling of nutrients from rocks and soil organic matter to plants. We also know that a plant's interaction with certain soil microbes influences the production of phytochemicals that bolster plant defenses and health—and the health of people and animals that eat them. These are among the basic mechanisms that Howard, Balfour, and Albrecht searched for to explain why agricultural practices that alter populations of microbes shepherding nutrients into plants may indirectly impact our health as well.

The level of nutrients, and especially minerals, in crops and livestock have long been of interest because insufficient levels translate into health problems for people. Nutritionists and geologists mean different things when they talk about minerals, though. While geologists consider minerals to be rock-forming crystals, like quartz, nutritionists refer to the individual elements that originate in rocks as minerals, or micronutrients. For now, we'll adopt the nutritionists' definition.

Micronutrients—like copper, magnesium, iron, and zinc—are among the elements essential for building phytochemicals, enzymes, and proteins in plants central to their health and the health of all who eat them. Micronutrient malnutrition is like a hidden hunger and now affects far more people than caloric malnutrition. Mineral deficiencies are estimated to afflict a third to half of humanity, causing major health problems in both developed and developing countries.

Researchers have connected mineral deficiencies to a wide range of physical and mental health problems in people. Copper is essential for hemoglobin to function properly and normal bone formation. Magnesium is a required element for at least three hundred enzyme reactions, and inadequate levels have been implicated in ADHD, bipolar disorder, depression, and schizophrenia. One recent study showed that depriving mice of dietary magnesium rapidly led to changes in gut microbiota and the development of systemic and intestinal inflammation. Iron deficiencies lead to anemia and reduce a person's capacity to learn or work. Zinc is required for at least two hundred enzyme reactions, and is essential for normal growth, tissue repair, and wound healing. Zinc deficiencies lead to increased susceptibility to infectious disease. These are only a few examples of the impact and importance of micronutrients.

You may need only a trace amount of micronutrients, but getting enough of them is critical to your health. It is only recently that we have turned to bottled pills and powders to get sufficient micronutrients. For most of our evolutionary sojourn we relied solely on foods in our diet. One of the longest-running analyses of the nutritional content of food began in 1927. Robert McCance, a doctor at King's College Hospital in London, initiated a study on the carbohydrate content of fruits and vegetables to help provide dietary guidance for diabetics. When postgraduate student Elsie Widdowson pointed out an error in his analysis of the sugar content of fruit, her sharp eyes led to a six-decade-long partnership investigating the chemical basis of nutrition and health. In 1940, they published *The Chemical Composition of Foods*, the first in a series of comprehensive analyses of common foods. Their work became the gold standard for dietitians and provides the most complete historical record of the mineral content of foods grown in the U.K. Regularly updated, it is still used today.

Over sixty years later, a curious nutritionist with a degree in geology and experience in mineral exploration compared McCance and Widdowson's 1940 edition with the 1991 and 2002 editions. David Thomas wanted to assess long-term trends in the concentrations of micronutrients in vegetables, fruit, meat, cheese, and dairy. It proved an interesting exercise.

In his comparison, Thomas found that, with the exception of phosphorus, significant losses in mineral content had occurred in each food group. Between 1978 and 1991, zinc levels in fruits and vegetables dropped by 27 percent and 59 percent, respectively. Averaged across all food groups, copper declined by 20 to 97 percent between 1940 and 1991. Magnesium dropped by up to 26 percent. Iron declined by 24 to 83 percent. When industry trade associations were quick to chalk up these differences to changes in analytical procedures, Thomas pointed out that, based on the description of the original scientific methodology, the laborious methods employed in 1940 were as accurate as modern automated methods—they just took longer. In addition, Thomas noted that the data from 1940 was taken from vegetables boiled for twice as long as those in 1991, which would have *reduced* the nutrient content of the earlier samples.

Of particular concern to Thomas was the startling decline in the mineral content of two staples of the English diet—potatoes and carrots. From 1940 to 1991, British spuds lost about a third of their magnesium, and almost half of their iron and copper. And carrots lost three-quarters of their magnesium and copper, and almost half of their iron. Two other crops, spinach and tomatoes, lost about 90 percent of their copper.

Though controversial, Thomas's study is not unique. Other researchers using different methods have also reported historical declines in food micronutrients. In one case, analyses of archived Kansas wheat samples showed that iron and zinc declined significantly from 1873 to 1995. And a 2004 study comparing nutrient levels in forty-three crops measured in 1999 to USDA benchmark nutrient studies from 1950 found significant declines in the median concentrations of protein, calcium, iron, phosphorus, and vitamins B2 and C. A subsequent 2009 review concluded that there was strong evidence for 5 percent to 40 percent declines in the mineral content of fruits and vegetables over the previous fifty to seventy years.

Agronomists generally attribute the decline in micronutrients over the past half-century to either depletion of these elements from the soil or to the "dilution effect." The former happens when nutrients taken up by crops are not returned to the fields. Over time the soil runs low in key nutrients if they are removed faster than they are replaced by new elements slowly weathering out of the rock. The latter refers to the situation with newer varieties of crops with larger or more numerous edible parts than older varieties. With more seeds per plant, in the case of cereal grains, or a bigger flower head, in the case of broccoli, the minerals in high-yield crops are spread through greater biomass and thus less concentrated. It's like spreading a tablespoon of peanut butter on a piece of bread instead of on a cracker. The peanut butter will be thicker on the cracker. Researchers point to evidence that implicates both depletion and dilution in the declining mineral content in crops following adoption of fertilizer-intensive agricultural practices.

In addition, food processing causes further mineral loss before crops become food. Milling and processing remove well over half of the iron and zinc from the world's main grain crops (as well as nutritionally valuable proteins and fats). All together, changes in agricultural practices and the nutrient losses due to milling and processing create a recipe for reduced nutrition—Albrecht's scenario of calorie-rich, nutrient-poor foods. But evidence is now emerging that another factor may be contributing to the historical decline in mineral nutrients in unprocessed foods.

DOUBLE-EDGED LEGACY

Conventional agriculture either directly or indirectly alters the microbial communities that deliver or influence the movement of micronutrients from soil to plants. Low iron and zinc are among the most common nutritional deficiencies in people despite levels high enough to support mineral-dense crops in most soils. In many cases, iron, zinc, and other micronutrients readily combine with other elements, like oxygen, to form relatively insoluble compounds. Though close by in the soil, they remain locked up and unavailable to plants. Certain microbes can help pry these elements loose. And this poses an overlooked question for modern agriculture. What if our practices knock essential microbial longshoremen off the job, stranding micronutrient cargo just offshore of the rhizosphere?

No one in mainstream agriculture worried about this problem. Indeed, by the early 1950s, widespread use of agrochemicals had significantly boosted yields and tamed many crop-damaging pathogens. With such miraculous results, their use increased, much like the growth of antibiotics in medicine during the same period. Who could argue with the use of chemicals that immediately delivered desirable results?

Biology, however, is mercilessly efficient when it comes to plants and animals expending energy because having it and getting it are central to survival. When fertilized, plants don't need to spend as much energy to get nutrients, so they don't grow as extensive a root system or produce as many exudates. This translates into fewer mycorrhizal fungi and beneficial bacteria in the rhizosphere. The net result is a decline in the nutrient exchange, mineral uptake, and phytochemical production vital to plant health and defense against pathogens.

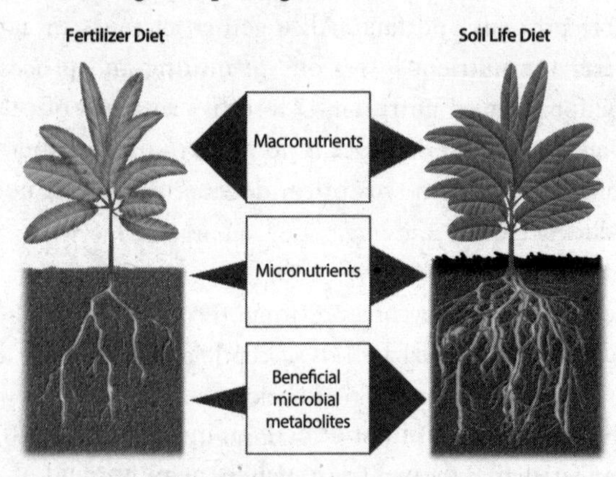

A Plant's Diet Matters. Soils high in organic matter support more diverse and abundant communities of soil life. Such communities enhance micronutrient availability and yield a profusion of compounds beneficial to plants.

Mycorrhizae are known to greatly enhance plant uptake of micronutrients. In one study, mycorrhizal fungi were found to deliver up to 80 percent of the phosphorus, 25 percent of the zinc, and 60 percent of the copper that made it into plants. And a 2004 comparison of wheat grown under organic and conventional practices in southeastern Australia found that conventional fertilization not only increased crop yields and phosphorus concentrations, it also reduced zinc uptake.

The cause? Dramatically reduced root colonization by mycorrhizal fungi. The fungi lost their reason to associate with plant roots, and the plants lost their micronutrient delivery service. This is why the conventionally grown wheat had less zinc than the organic wheat grown in neighboring fields. Fewer exudates meant less zinc. This is the kind of mechanism Howard hypothesized but could not prove. And, much as he would have expected, a 1988 study on fields in India found that additions of organic matter increased zinc mobility and plant uptake.

Could changes in agricultural practices over the past half-century have reduced populations of micronutrient scavenging and delivering bacteria and mycorrhizal fungi enough to affect mineral uptake of crops? The idea remains controversial, and any such effects are likely to be highly variable depending on soil type, crop varieties, specific farming practices, and legacies of past farming methods. But still, given what scientists have been uncovering about the exchanges and communication in the rhizosphere, it would be surprising if there were no effects.

It's worth noting that the way we actually work the soil has an impact too. Conventional plowing influences the composition of soil microbial communities. No-till farming, which ditches the plow in favor of drilling seeds into the ground, leaves the plant litter from the prior crop on the field, thereby reducing erosion and increasing soil organic matter. This practice also considerably reduces the physical impacts on soil life and increases the abundance of beneficial mycorrhizal fungi. Perhaps this is why potatoes grown under no-till farming are less susceptible to pathogenic fungi than those grown in conventionally plowed fields.

Even with the benefits of no-till farming, the application of copious amounts of fertilizers, pesticides, and herbicides remains problematic because they disrupt soil biology. These chemical products change the composition of communities of bacteria and mycorrhizal fungi, impacting symbiotic relationships in the rhizosphere that influence the movement of micronutrients. While there will be microbes in the soil no matter what we do (short of sterilizing it), the real question is which ones will be dominant—those that work for us or those that work against us?

Recent studies confirm that organic agriculture promotes greater beneficial microbial biomass in the rhizosphere. A field study in Brazil found that converting from conventional to organic farming of acerola (better known as the Barbados or West Indian cherry) increased microbial biomass 100 percent to 300 percent over a two-year period. Similarly, a two-year field test in North Carolina comparing the effect of organic and conventional tomato farming found that the application of composted cotton gin trash to organic fields more than doubled both soil microbial biomass and the amount of plant-available nitrogen compared to fields fertilized with commercial synthetic fertilizer.[1] On top of this, straw mulching increased soil microbial biomass and plant-available nitrogen by an additional 43 percent and 30 percent, respectively. Another conversion study, from conventional to an organic mixed-crop rotation on an experimental field in North Carolina found substantial increases in microbial biomass. After yields declined in the first two years of the study, the productivity of the organic plots caught up with or exceeded the conventional plots by the third year.

How did this work? A greater amount of soil life increased the nitrogen reservoir available to plants. Consider, for example, how nematodes that eat soil bacteria excrete nitrogen-rich waste back into the soil, thereby turning bacterial biomass into plant-available nitrogen—organic fertilizer. It seems as though Sir Albert Howard was on the right track after all.

A long-term agricultural experiment established in Switzerland in 1978 found that organic practices and the application of farmyard manure did not just increase microbial biomass, soil carbon, and soil nitrogen content. There were half as many aphids in organically fertilized fields due to a twofold increase in the number of predatory spiders. In other words, the increase in soil bacteria populations rippled up from a below-ground ecosystem to support predators that served as biological pest control for the crops above ground.

Scientists at the University of Illinois recently proposed that intensive nitrogen fertilization stimulates microbes to rapidly degrade soil organic matter, thereby depleting it as a reservoir of nutrients. Through analyzing data from the world's longest-running experimental corn plots, they found significant declines in the amount of carbon and nitrogen in soil organic matter despite leaving substantial organic matter (crop stubble) on the plots and applying liberal amounts of nitrogen. Their findings were unexpected, as it had long been assumed that adding synthetic nitrogen helped build soil organic matter. It was thought that growing more plants on a given piece of ground would provide for more organic matter by leaving more crop stubble after harvest. The conclusion that nitrogen fertilization instead accelerates the breakdown of soil organic matter upends conventional wisdom.

For decades, few have been concerned about the steroid-like effects of nitrogen fertilizer on microbes, despite troubling studies that predated institutionalization of fertilizer-intensive agriculture. In 1928, Selman Waksman, the soil scientist turned antibiotic hunter we met earlier, documented that the addition of inorganic nitrogen, phosphorus, and potassium more than tripled the pace of microbial breakdown of soil organic matter. And a decade later, in 1939, William Albrecht measured organic matter levels in experimental plots after fifty years of fertilizer treatments. He too found that the addition of inorganic fertilizers reduced soil organic matter. But no one much cared then, given the remarkable gains in yields that fertilizers produced.

Howard's fears about a growing addiction to agrochemicals now appear well founded. Once soil organic matter is degraded, fertilizer becomes essential to maintain yields. Yet fertilizers are only part of the problem. Pesticides and herbicides have also altered soil microbiota in ways we do not fully understand. Some studies, however, point to effects that echo the basic problem with the Western diet—like overconsuming refined carbohydrates, excessive use of agrochemicals feeds the bad actors and starves the good ones.

It's not surprising that agrochemical manufacturers strenuously challenge research that points to the negative effects of biocides—herbicides, bactericides, fungicides, and insecticides. With one for each major branch on the tree of life, they have a lot riding on the issue. Consider, for example, the case of the herbicide glyphosate. Early studies found that glyphosate had low direct toxicity to humans and rapidly broke down. But, scientists report in several recent studies that glyphosate may not be so innocuous after all. The authors conclude that the negative effects are due to the disruption of microbial communities rather than acute toxicity. These studies include experiments that show glyphosate affects rhizosphere microbiota, reducing plant uptake of nutrients (like phosphorus, zinc, and manganese). Researchers also found that glyphosate altered the gut biota of poultry and cows, which allowed pathogens to proliferate at the expense of beneficial bacteria. In light of this, a 2014 paper in the journal *Food Chemistry* that reported significant levels of glyphosate in marketable soybeans will no doubt spur further questions about the bestselling herbicide in the world.

MICROMANURE

Perennial controversy surrounds comparisons of the nutritional composition of organically grown foods and those grown conventionally. Yet a clear picture emerges when you ask the question a different way—what happens if soil microbes known to deliver micronutrients to plants are added to soil? Mineral uptake increases, sometimes by a lot. The beneficial effects of inoculating soil and plant roots with certain bacteria and mycorrhizae are as incontrovertible as the effects of known plant pathogens. We now know that root exudates influence the composition of rhizosphere microbial communities—and are beginning to understand how microbial metabolites stimulate plants to produce phytochemicals with antibacterial and antifungal properties. This opens the door for manipulating rhizosphere microbiota to suppress pathogens, and thereby promote plant health. Indeed, microbial partnerships hold tremendous, and greatly overlooked, potential to reduce or replace pesticides and fertilizers and help sustain intensive agriculture. Herein lie parallels to the promise of probiotics and prebiotics in medicine.

Bacterial inoculation of plants is like dietary probiotics in that both introduce live microbes that benefit the host. Laboratory studies and field trials have shown that inoculating the root microbiome with the right bacteria can promote growth, enhance nutrient uptake, help control pathogens, and induce increased root surface area and shoot length. There is even an acronym for bacteria that can spur plant growth: PGPR—plant-growth-promoting rhizobacteria. Those that increase nutrient availability and uptake are called biofertilizers. The growing understanding of the many ways in which the plant microbiome influences plant health offers promising opportunities to enhance crop quality and yields.

Inoculating plants with beneficial bacteria can help protect against pathogenic fungi and bacteria. In one example, inoculating eggplant with symbiotic bacteria reduced wilt caused by pathogenic bacteria by 70 percent. And studies in sugarcane and rice have shown that inoculation with beneficial bacteria increased the expression of plant-defense-related genes. In other words, applications of the right kinds of bacteria prepare plants to repel pathogens through teeing up their defenses the way commensal gut bacteria help keep the immune system poised to repel pathogens.

The takeaway here is that chemical signaling across the root surface is central to plant defense. Stuck in place, a plant can't move to avoid pests and pathogens. And it would be inefficient for plants to continuously secrete antimicrobial compounds when they are not constantly threatened. But it makes a lot of sense for them to exude compounds tailored to recruit allies. Then, when under attack, plants can signal bacteria to produce defensive compounds directed at the bothersome pest or pathogen.

Many farmers already apply cultured strains of specific rhizobia to their crops, either as a coating on seeds or sprayed on the fields at planting, to increase nutrient uptake and boost crop yields. Studies over the past several decades have shown that introducing different types of beneficial bacteria can significantly increase plant growth and crop yields for the big three grains—wheat, corn, and rice—as well as for barley, canola, sorghum, potatoes, peanuts, and various vegetables (lettuce, tomatoes, peppers, peas, and beans) and tree crops (apples and citrus). Among these studies, field trials on wheat reported yield increases of 10 to 43 percent. Experiments with both greenhouse-raised and field-grown corn (*Zea mays*) showed that inoculating seeds with bacteria possessing phosphate-solubilizing abilities and other plant-growth-promoting traits increased yields as much as 64 to 85 percent compared to uninoculated plants. A 2009 study on corn reported that plant-growth-promoting bacteria could cut the use of phosphorus fertilizers by half with no loss in yield. In another study, additional inoculation of wheat with strains of *Bacillus* increased yields by up to 39 percent over fertilizing with just rock phosphate. Biofertilization offers one of the few real options for rapidly increasing crop yields.

While commercialization of bacterial inoculants has been hampered by limited understanding of rhizosphere ecology, it is clear that one way they spur crop growth is through increasing plant uptake of phosphorus, the availability of which limits crop growth on nearly half of the world's arable land. Phosphorus is the least abundant macronutrient and is the second-most-important limit on crop growth (after nitrogen). Most of the world's phosphorus reserves are held in certain types of phosphate-rich rocks and in relatively rare minerals. In the soil, phosphorus readily forms insoluble compounds with calcium, iron, and aluminum, rendering it unavailable to plants. Certain bacteria, however, can liberate the locked-up phosphorus, converting it to the soluble form that plants can take up.

And microbes also help plants access phosphorus from another source—organic matter. The remains of dead plants and animals often account for half of the total phosphorus in soils and sometimes as much as 95 percent of it. Microbes that break down soil organic matter release phosphorus and other mineral-derived elements back into biological circulation. It's no mystery why the rhizosphere generally contains far higher concentrations of bacteria that can solubilize phosphate than the soil at large. Phosphate-solubilizing bacteria flock to exudates, bringing goods to trade for sugar.

Using a radioactive isotope of phosphorus as a tracer, scientists have confirmed that phosphate-solubilizing bacteria also partner with mycorrhizal fungi. Bacteria consume stable phosphorus-bearing compounds and release soluble phosphate into the soil, where fungi deliver them to plants. In this case, the mycorrhizal fungi serve as middlemen. Here again is another mechanism that Sir Albert Howard intuited, one that shows the potential for designing microbial cocktails to spur the growth of particular crops in particular soils. Instead of the approach pursued for the past century—dumping enough phosphate into the soil to ensure some gets to crops—perhaps we can employ microbes to shake loose the tightly bound inaccessible phosphorus that is already present in agricultural soils.

By some estimates, global reserves of plant-available soil phosphorus will be used up by 2050. In a world where increases in crop yields have stalled out, and the population continues to increase, using microbial inoculants to liberate phosphorus and spur crop growth may prove crucial to feeding the hungry world of tomorrow. Cuba, for example, began to commercially produce biofertilizers when their fossil fuel supplies were significantly curtailed after the collapse of the Soviet Union. Lessons from their experience could matter greatly in the future, as the accumulated amount of phosphorus bound up in the world's agricultural soils is estimated to be sufficient to sustain crop yields for a century—if, that is, plants can access it.

Microbes can even help us mine phosphorous from our own waste, something we'll need to do eventually given the limited amount of phosphorus and the growing number of people on the planet. Certain species of *Acinetobacter* bacteria are like vacuums, sucking up enough phosphate to account for as much as 80 percent of their total weight. Environmental engineers can grow populations of these phosphate-scarfing bacteria by passing raw sewage through two tanks, the first anaerobic and the second aerobic. Called enhanced biological phosphate removal, this process efficiently concentrates and removes phosphorus from sewage. Similarly, nitrates can be microbially scavenged during sewage treatment, a process called bacterial denitrification. If we give microbes the right conditions to work under, they will clean up and recycle our waste, producing fertilizer. Here then is symbiosis industrialized—both we and the environment win.

Rather than genetically engineering plants to tolerate new agrochemicals with unknown side effects, it makes as much (if not more) sense to support the beneficial relationships between plants and microbes that underpin plant health, and thereby agricultural productivity. Biofertilizers are already conservatively thought to be capable of replacing a quarter to a third of conventional fertilizers.

In addition to capitalizing on what microbes already do, harnessing their capabilities in new ways may offer other promising approaches. For example, there is a substantial interest in creating symbiotic associations between nonlegume crops, like wheat and corn, and nitrogen-fixing bacteria to reduce the demand for nitrogen fertilizers. Understandably, conventional fertilizer companies are not thrilled with this idea. But could such a breakthrough precipitate a refreshing alliance between advocates of sustainable agriculture and those who promote genetically engineered crops?

Of course, the simplest solution for restoring fertility to the soil is to follow a prebiotic strategy and feed the beneficial microbes. In an agricultural context, the food for microbes is called biotic fertilizer (not to be confused with biofertilizers, which are live microbes analogous to probiotics). Nature relies on prebiotics—organic matter, compost, and mulch—to build fertile soil. But simply returning organic matter to agricultural fields (think mulching) isn't easy to do on a large scale, which is why using certain prebiotics in the soil is so compelling. These biotic fertilizers work like organic matter—they feed soil life—only much faster. Advocates trumpet that soil fertility increases dramatically in just a couple of years.

A particularly intriguing variant of the biotic-fertilizer approach promotes the growth of certain types of cyanobacteria (blue-green algae) that live in soil. What's so special about cyanobacteria? They can photosynthesize *and* fix atmospheric nitrogen. How well do they work? For centuries before chemical fertilizers, naturally abundant populations of cyanobacteria helped keep rice paddies fertile across much of Asia.

Feeding cyanobacteria creates an ecological ripple effect. The initial bloom of cyanobacteria expands biological activity and jump-starts soil fertility many times faster than when nature is left to her own devices. As cyanobacteria and other microbes die, they rapidly break down, increasing soil carbon, nitrogen, and plant-available nutrients. The basic idea is that cyanobacteria trigger the growth of secondary microorganisms, thereby providing sufficient organic matter to keep the process going until the next biotic fertilizer application.

Consider, for example, one product commercially available in Washington State. The manufacturer collects chicken manure from major feedlots and uses a proprietary composting method to kill pathogens and break down the antibiotics and pharmaceuticals fed to the chickens. The composted product is then converted into a concentrated fertilizer blended to promote the growth of cyanobacteria. Farmers who used this biotic fertilizer report that their crops became more resilient to fungal and insect pests, and that mineral uptake increased. Though introduced for use on organic farms, biotic fertilizers are finding increasing use on conventional commercial farms.

With the growing interest in reducing the need for nitrogen fertilizers, inoculating soil with beneficial bacteria and biotic fertilizers offers a new strategy. Like the darkening soil in our garden, soils that repeatedly receive biotic fertilizer applications see their organic matter content increase, in some cases by more than 1 percent a year. Proponents of this method caution that biotic fertilizers need to be tailored to the specific cyanobacteria in the soil and a region's particular environmental conditions (e.g., temperate or tropical, wet or dry).

While the potential for using microbial ecology to enhance soil and crop health and yields appears promising, recognition of a biological basis for soil fertility still faces formidable obstacles. After all, modern agricultural practices are steeped in a century and a half of chemically centered theory and practice. And on a practical level we are also limited by our embryonic understanding of microbial ecology. Yet that is changing—rapidly. The growing commercial appeal of biofertilizers reflects the fact that some approaches are already cost-competitive with conventional fertilizers and can produce comparable, if not greater, yields. Chemical fertilizers will, no doubt, remain integral to commercial agriculture well into the future. But we'd serve our descendants well by starting to cut back on overapplying them and to focus on rebuilding soil organic matter—nature's time-tested foundation for perpetually fertile soil.

INVISIBLE FRONTIERS

Microbes living inside and around plant roots are as essential to the defense system of plants as our microbiome is to our immune system. People are dialed into the same biologically based defense strategy as plants. We both offer up nutrients that attract microbes to a specialized zone—the rhizosphere in the case of plants and the colon in our case. These places serve as biological bazaars where microbes trade nutritional wares and form alliances with plants and people.

One of the papers we came across while researching this book described gut microbes feasting on exudates in the mucosal lining of colon cells. Exudates in the colon? Weren't exudates part of the botanical world? Then it clicked. The root is the gut and the gut is the root!

It's probably not a coincidence that the majority of the bacteria in the gut and in the soil share a saprophyte genealogy (from Greek, *sapro* refers to rotting things and *phyte* means "plant"). In both places, the bacteria that are present specialize in decomposing dead plant matter.

If you were to turn a plant root inside out, rhizosphere and all, you would see that it is like the digestive tract. The two are, in many respects, parallel universes. The biology and processes that bind soil, roots, and rhizosphere together mirror those in the mucosal lining of the gut and the associated immune tissue. The gut is the human version of the rhizosphere, the part of our bodies incredibly rich with specially recruited microbes. While cells of the digestive tract interact with gut microbes, root cells cut deals with soil microbes. The human world and the botanical world share a common theme—lots of communication and exchanges with microbes.

But the common thread linking the gut and the root runs deeper yet. Our teeth do the work of soil detritivores, chomping and chewing away at organic matter to make it smaller, which allows other organisms to continue the decomposition process. Stomach acids function like the fungal acids in soils, breaking food down into absorbable molecules. The small intestine absorbs nutrients much the way plant roots absorb nutrients dissolved in water. And the inside of the small intestine is carpeted with small, thread-like projections called microvilli that increase the surface area many times over, greatly enhancing nutrient absorption—just like root hairs in the soil. In the cauldron of the colon, as in the rhizosphere, microbes make some of the most critical metabolites and compounds that their host needs.

Goblet cells found in the wall of the small intestine and colon produce a thick layer of mucus that protects the rest of the cells and keeps the luminal contents moving along. At one time, scientists thought this was the sole reason your colon produced mucus. Then they discovered that bacteria live in and eat the mucus. It's like the carbohydrate-rich exudates that plants release at the root cell surface to feed rhizosphere-dwelling microbes. Legions of bacteria in your inner soil feast on mucus as well as undigested plant matter and dead colon cells. In return, their metabolites nourish your colon and their presence helps deter pathogens.[2] The way that our microbial partners use the raw material of our diet to manufacture an arsenal of beneficial and defensive compounds mirrors the interactions between rhizosphere microbiota and plant roots.

Soil life that breaks down organic matter produces a steady flow of nutrients to plants. This echoes the way bacteria in the colon convert complex carbohydrates into a steady supply of beneficial compounds (like SCFAs). In each case, a diet rich in plant organic matter delivers key nutrients critical for health and well-being. Simple carbohydrates and single mineral fertilizers, on the other hand, fuel rapid growth but they do not supply the full range of nutrients that underpin the health of plants—or the people who eat them.

While a root thrusts itself into the world of soil to find food, we bring the outside environment directly into our body. This is ultrarisky considering all the things in the outside environment that might (and do) harm us.

Our gut, like a plant root, must sieve through the different materials it encounters—what we eat and drink—separating food from foes and friends. The border between the gut and its contents, and between the root and the soil, defines invisible frontiers across which all nutrients must pass. Microbes are the middlemen—the planet's tiniest truckers. Chemistry defines the cargo that moves across these frontiers, but biology keeps the commerce at the root of life buzzing.

If you were to take a close look at a cross-section of your colon, you'd see bacterial cells shoulder-to-shoulder with your own cells, making it hard to say where they end and you begin. Consider for a moment, if you will, that beneficial bacteria are likely tucked away in your colonic crypts. Consider that this picture looks pretty similar for a plant, given the way mycorrhizal fungi enter a plant root and squeeze in between the cells. The evolutionary adaptations between microbes and plants—and between microbes and our bodies—are as astounding as they are intricate. In both cases, the quality of the soil at the invisible frontiers of root and gut, and whether we poison, neglect, or cultivate it, are central to the health of plants and people.

The striking parallels between the roles of the respective microbiota of the root and the gut suggest connections that are both fundamental and universal. In both places, a community of microbes assists their host with two of life's essentials—obtaining food and defending against foes. In return, they get some of the best habitat a microbe could hope for, safe space rich with regular sustenance—an ideal place in which to reproduce and live the microbial good life.

Whether in plants or ourselves, evolution has favored microbes that develop partnerships benefiting both host and microbial tenant. This is likely no coincidence. We harbored symbionts ever since we came into being. Some of the same species that inhabit the inner soil of our guts live in soils, where they help suppress plant diseases. This simple, and until recently unappreciated, reality lies at the root of how to begin retooling agricultural and medical practices—by courting ancient friends.

14

CULTIVATING HEALTH

We can't help but see the world differently after unearthing the parallels in the essential roles that microbes play in both soil health and human health. While we still can't see the half of nature hidden beneath our feet, we know it is the root of the life and beauty we see in our garden every day. And we look at ourselves differently too knowing that we are each a tribe of trillions.

Awed by the realization that the animals, plants, and landscapes we see around us are merely the visible tip of nature's iceberg, we now appreciate how the mysterious world of microbes helps make soil fertile and food nutritious. We had thought most microbes were harmful, foes for our immune system and antibiotics to vanquish. Yet microbial communities are integral to key aspects of our own metabolism. Learning that we reap the harvest of what we feed our soil—inner and outer, for better and worse—widened our view, bringing into focus the extraordinary agricultural and medical value of cultivating beneficial microbes in the soil and in ourselves.

For well over a century humanity has viewed our invisible neighbors as threats. We saw soil life primarily as agricultural pests, and through the lens of germ theory we typecast microbes as agents of death and disease. The solutions that grew from these views—agrochemicals to eradicate pests and antibiotics to kill pathogens—became embedded in our practices. Intent upon killing bad microbes, we haven't cared much about the collateral damage to innocent microbial bystanders, although we are beginning to glimpse the effects upon ourselves.

While spraying broad-spectrum biocides on fields may take care of agricultural pests over the short run, the pests return with a vengeance in the long run. And there is a direct parallel to aggressive use of antibiotics over recent decades, which have spawned new strains of antibiotic-resistant bacteria, an increasing number of which we now have no defense against. Instead of solving problems, we've become addicted to solutions with limited staying power. Dowsing gardens, farms, and people with broad-spectrum biocides should no longer be the de facto solution for gardeners, farmers, and doctors.

What does this all mean? Soil fertility and our immune system—two things critically important to us all—don't work like we thought they did. Plants with depauperate communities of beneficial microbes in their rhizosphere dial back producing the phytochemicals that defend them and nourish us. Of particular relevance for our own health is that it turns out we need most of the microbes we've been trying so hard to kill. And scrambling our own microbiome, especially early in life, is increasingly implicated as a factor underlying modern maladies. It's not that we shouldn't fight pests and pathogens, but that the approaches we have come to rely upon come with hidden costs.

Looking back on our experience, we believe the difference between a garden and a weed-covered lot can show the way forward. Nature abhors bare ground, and she'll fill it in her own way. But you can shape a place if you work with her. We intentionally cultivated the soil beneath our tiny patch of Earth to reap flashes of color from flowers, trees to inspire us, and vegetables to eat. The discovery that the real source of the beauty, comfort, and sustenance in our garden lay beneath our feet surprised us, and so did something else about our garden. It has about the same surface area as our digestive tract. Imagine gardening your gut, tending to the life you want and need in the body's innermost sanctum.

Just as compost, wood chips, and mulch nourish soil life, the same is true of the foods that nourish the symbiotic inhabitants of our gut. While a living soil will ripple above ground to support the health and resilience of a garden or farm, your inner soil supports another kind of garden—your body. If we cultivate the microbes that benefit us, they'll help fend off their pathogenic cousins and keep our immune system working for us, rather than turning against us.

Tending the garden of our microbiome doesn't mean forgoing modern medicine. Realistically though, it's going to take some time to align medical practices and therapies so that they work with our microbiome. In the meantime, we need to ensure we start out with a healthy microbiome and then maintain it with a diet rich in prebiotics. And if our microbiota take a hit, whether after antibiotics, illness, or maybe even a colonoscopy, we might consider doing what a gardener does and replant what we've lost and help them get established.

In the end, it boils down to some simple advice. Starve your enemies and feed your friends. And don't kill off your allies that help keep the enemies in check.

Although we have but a hazy view of the full scope of the microbial ecosystems that are our bodies, we already know enough to start changing certain practices. Most obvious is prescribing antibiotics with a lot more thought and care for our children, ourselves, and our farm animals. The same goes for easing up on obsessively sterilizing our own homes and bodies. And while restoring the microbial ecosystem of the gut sounds like a tall order when we are just learning what species live there and how they interact, we might consider a key lesson from the natural ecosystems we know. Degraded ecosystems are notoriously difficult—and expensive—to restore. Preventing harm in the first place is generally the best long-term strategy.

Van Leeuwenhoek was onto something big with his tiny curiosities—he just didn't know how big. Over the centuries that followed, pioneering souls illuminated the mysterious dark side of the microbial world. In the process, they defeated many of the worst human pathogens. But just when we thought we had microbes pegged as our invisible enemies, early-twentieth-century visionaries glimpsed the beneficial roles microbial life plays around us and within us. By the close of the century, scientists had pulled the curtain back enough to reveal the startling microbial lineage of higher life. Today, we are just beginning to see how our entangled alliances with microbes are redefining who we are and what we're really made of.

While there have been astounding advances in applied microbiology over the past half-century, most progress focused on pathogenic microbes we could culture. Our success in controlling many infectious diseases constrained our thinking and practices in the way a lantern defines a circle of light. Looking in the illuminated circle, we didn't much consider what lay beyond—the ecological interactions among communities of hard-to-culture microbes and their role in our well-being. Now that gene-sequencing technology lets us light up the shadows, we are seeing much more of the microbial world—and what microbes do.

What we're discovering is that our microbiome and those of plants hover at remarkable, and remarkably similar, frontiers. From the rhizosphere to the mucosal lining of the colon, microbes thrive at interfaces. They always have, ever since the first stromatolites set up shop on the shores of ancient seas. Such environmental borderlands offer a platform on which a microbe can live close to a steady supply of nutrients. Along their evolutionary journey, some microbes threw their lot in with plants and animals, colonizing root surfaces and gut walls, and helping to screen out what is harmful, usher in critical nutrients, exchange information, and pass on important metabolites to their hosts. In so doing, the smallest creatures that ever lived helped keep plants, animals, and all of our ancestors cruising along for millions of years right on up to today.

We were surprised to learn that the environmental systems on which we depend are founded on cooperation as much as on competition. Symbiotic relationships are not the outliers that textbooks portray. Diversity nested in cooperation creates dynamic systems that can stand the test of time. And though scientists may never know all the mechanisms and ways in which these complex relationships work, new studies continue to show the potential power of microbial symbioses in agriculture and medicine.

We believe the revolutionary advances in microbiome science will continue reshaping scientific understanding of nature's hidden half at many levels—from our bodies and yards to our neighborhoods, cities, farms, and forests. The growing awareness that we and all plants and animals evolved along with our microbiomes is helping to crystallize new views of the natural world and our place in it. Views so radically different that when we cracked open our college biology textbooks they had nothing much to say about our microbial side. Advances in plant and human microbiome research over the past two decades have upended and reshaped the understanding of biology that most mid-career scientists, doctors, and farmers learned in college. Now it seems that, just like infectious diseases, the current barrage of autoimmune and chronic diseases may also have microbial roots.

Microbiota vary greatly from person to person and day to day within a person, and both genes and environmental factors are sure to figure into how microbes influence autoimmune and chronic diseases. Because of such complexities, microbiome researchers rightly caution against overselling the exciting new discoveries in their field. And it would be easy to do so given the astounding connections that are coming to light.

Indeed, dysbiosis, the opposite of symbiosis, is now under investigation as a primary contributing factor to a long list of maladies. Among these ailments are leaky gut syndrome and inflammatory bowel disease, as well as obesity, certain cancers, asthma, allergies, autism, cardiovascular disease, type 1 and type 2 diabetes, depression, and multiple sclerosis. Where associations and emerging causal linkages between dysbioses and disease will lead us, no one can be sure. Still, it is clear that exploring the microbiome is opening doors to potential treatments and cures for many modern sicknesses—including ways to kick our agrochemical habit.

Imagine the day in which a survey of your gut microbiome will become an index of your personal health, along with your temperature and blood pressure. Likewise, tailoring soil microbes to different crops grown in different soils in various regions and climates may become a key tenet of sustainable agriculture. We are not there yet in either case, however, as we still have much to learn about conducting and interpreting such a survey.

Although it will take time to develop new practices, what we are learning about the microbiome has fundamental implications for agriculture and medicine today. You don't have to be a researcher in microbiology, immunology, or plant science to see the new paradigm—cultivate the good guys in the soil beneath your feet and in your gut's innermost sanctum.

Learning to work with our ancient microbial friends means using long-term thinking to guide short-term practices—something easy enough in theory, but much harder to do. It is difficult to let go of beliefs, especially those reinforced by our parents, advertisers, and society as a whole. From the time we are small, we are told to not play in the dirt and to mind the five-second rule. Shop for nearly anything and it's easy to see how thoroughly germ theory has seeped into our lives. We are encouraged to coat our hands and bodies in antimicrobial products and sanitize our world with all manner of disinfecting products. Antimicrobial chemicals are embedded in plastics, shoe liners, clothing, toys, TV remote controls, keyboards, and steering wheels. This is not to say we should forgo rational hygiene—after all, Dr. Semmelweis demonstrated the wisdom of hand-washing a long time ago.

But while we've become incredibly successful at keeping many pathogens at bay, we're now realizing that practices built upon the foundation of germ theory can harm or destroy the beneficial microbes that reside in our fields and in our bodies. Creatures like bacteria—which swap genes like handshakes, reproduce at a furious pace, and can eat just about anything—will quickly overrun a blank slate. The fact that it is useful—and sometimes essential—to kill microbes doesn't change the reality that the indiscriminant use of biocides can disrupt or decimate beneficial communities of microbes.

As it becomes clearer that dysbioses undermine our health and the health of the land, microbiome science is also revealing the time-tested basis for traditional practices and diets. For decades mainstream science dismissed such practices as the uninformed fruit of ignorance and superstition. And while this seems a fair assessment of certain practices, like the infamous red cure for smallpox, it is not true across the board.

We are now learning why some traditional approaches to plant health and human health work—they support the beneficial microbiota at the heart of the symbiotic relationships in the soil and our bodies. This is why it matters whether soil biota get enough organic matter to eat—and why it's worth eating the kind of carbohydrates that keep the alchemical cauldron of the colon bubbling with life. As odd as it might have sounded at the start of this book, mulching your inner soil can mean the difference between robust health and ill health.

The folly of practices that harm beneficial microbes haunts us, creating new problems without solving the original ones. This is the hallmark of a bad strategy. Burning up soil organic matter and starving beneficial soil life have left us with a legacy of barren fields. And likewise a diet poor in plants and rich in antimicrobials threatens our inner soil. For too long we have tried to replace biology with chemical nutrients and poisons.

Resurgent agricultural pests, declining soil fertility, crisis-level antibiotic resistance, and life-sapping chronic diseases all seem unrelated until you consider their roots in disrupted microbial ecology. Bacteria began resisting antibiotics almost from the moment we stood back and marveled at the killing power of these new miracle drugs. The more we try to poison bacteria, the more resistant they become because of their rapid generational turnover and ability to casually pass on the traits that shield them. We may win battles against them, but we will never win the war this way. We need a different strategy.

The root of a new plan for health in people, plants, and animals lies in recognizing that we have long lived in an ecological embrace with microbes, fine-tuning our relationships with them so that they help us run the internal environments of our bodies. We spent more than 95 percent of our several-hundred-thousand-year existence immersed in nature. Hunting and gathering wild foods and migrating across new landscapes brought us into contact with microbes that coated our bodies inside and out. This constant exposure to microbial life trained and toned our immune system. Then, in but a blink of Earth's eye, we felled forests, poisoned fields, and paved the land, depleting the stores of nature we once drew upon for our microbiome. In an evolutionary heartbeat, we began renegotiating partnerships honed over countless generations.

Today's agricultural and medical technology is stunning indeed. We can splice genes into plants to create instant evolution and send robotic tractors out to work the fields of enormous farms. We laser-sculpt our eyeballs so we can shed our glasses, and we transfer organs from one body to another. But we are only beginning to untangle the relationships among the inhabitants of microbial ecosystems. We're right back with Linnaeus, figuring out who is there and what to call them.

No doubt surprises await us. Indeed, as we were finishing this book, a new study showed that artificial sweeteners create dysbiosis in ways that alter how both mice and people metabolize glucose. It appears that sweeteners used in so-called diet sodas mimic the sugar they replace in some respects. Like most people, we had thought that calorie-free artificial sweeteners were useful for those watching their weight. But apparently our microbiota consider them much like sugar, leaving us all to wonder if artificial sweeteners are a back door to type 2 diabetes and obesity.

And the idea of going with your gut is taking on new meaning. Who would have thought that the inhabitants of Metchnikoff's colonic garbage can make serotonin, a neurotransmitter that shapes one's mood. Not only are our gut microbes communicating with our nervous system, but our emotional state can influence our gut microbes—and the spectrum of metabolites they produce.

Maybe we shouldn't be all that surprised when nature's hidden half throws us such curveballs. Aldo Leopold, an iconic figure well known to ecologists, watched what happened to the vegetation in the Southwest when people decided that killing off wolves would be a good idea. The deer population exploded and consumed the forest, leaving the land denuded and the deer without enough to eat. Little did trigger-happy ranchers and wildlife agents suspect that shooting carnivores would erode the soil and starve the deer.

What does this all mean for the way we treat microbes? It means we must find new ways to protect ourselves, our crops, and our livestock from pests and microbial pathogens while cultivating our allies. We need to blend the mind of an ecologist with the care of a gardener and the skill of a doctor. For working with the hidden half of nature points the way to some surprising—and surprisingly effective—ways to help address a wide range of seemingly disconnected environmental and health problems.

A couple of decades ago, it would have sounded crazy to argue that plants and microbes in the soil run a biological barter system that functions as a plant's defense system and allows us to harvest nutrient-laden plant foods essential to our health. Even more unbelievable would have been the notion that bacteria communicate with our immune system, helping it to precisely mete out inflammation to repel pathogens and recruit helpful commensals. These surprising new truths carry fundamental implications for the way we view, and should treat, a wide range of seemingly unrelated maladies. In medicine, as in agriculture, what we feed our soils—inner and outer—offers a prescription for health forged on the anvil of geologic time.

Anyone who walks with open eyes through gullied fields, concrete-lined streams, and stump-covered hills can see how our hands triggered these impoverished conditions. But it's harder to see how our actions change microbial landscapes—until we connect the dots and recognize the effects that manifest within us and around us. Changing our thinking about microbes is the first step toward changing how we see them. After all, while sight is an ability, seeing remains an art.

It is difficult to talk about, let alone act upon, conserving and protecting something you can't see. But if we are to do this, we need to see the world as it is, imagine what we want it to be like, and do what it takes to get there—instead of following our well-worn path of imagining how things are, ignoring how they work, and doing whatever we please. It is time to accept that our reality plays out under the invisible influence of microbial communities that keep nature's grand design—and our own self-centered lives—humming along.

Still, the vast majority of microbial species remain elusive to science, as do their relationships with one another and ourselves. Microbes serve as nature's software—an organic operating system alive with biotic code, a menu of genetic instructions crafted over deep time. Living in the background, keeping essential systems working, microbes shaped our ancestors' earliest days and continue to run the world we know.

Like most software, the microbial code is out of sight and out of mind—until it crashes, an error message pops up, or systems that once worked well start to fail. And it's no secret that software errors are hard to fix if you don't have the source code. We're only beginning to understand the language of microbial ecology and the biological programming built over the long haul of evolution. So perhaps we should think twice about doing away with code we don't understand. Beta testing new configurations of critical systems with neither blueprint nor backup plan is always risky business.

So where does this revolutionary new perspective leave us? Put bluntly, many practices at the heart of modern agriculture and medicine—two arenas of applied science critical to human health and well-being—are simply on the wrong path. We need to learn how to work with rather than against the microbial communities that underpin the health of plants and people.

For agriculture this means treating soil like what it really is—the living foundation of all life. To grow anything one must feed it, and the way to maintain fertile agricultural soil is to cultivate soil life with organic matter. Much the same idea applies to our inner soil. What we eat feeds and shapes our microbiome's metabolism, which in turn shapes our health—from the inside out, for better or worse. Of course, changing one's diet will not cure acute illnesses. But it may be the single most effective thing a person can do to prevent chronic illnesses and promote their overall health.

While scientists and doctors in a wide range of disciplines are poised in the decades ahead for discoveries that will inform new practices and therapies, some things are already pretty clear—things we can take action on now. This means eating for your microbiome, the living roots of your immune system. When your gut microbiota get their fill of complex carbohydrates, you harvest health.

When we bought our house, it had a decades-old kitchen split between two small rooms. We thought that remodeling the kitchen would improve our lives. It did, but in the end, remodeling the yard to make a garden truly changed us. Watching a place you thought you knew well slowly transform can help you notice the things that truly are right in front of you. Over time, we both came to see that what was outside the house, hidden beneath our feet, was as essential to our happiness and well-being as what was inside.

Making a garden taught us things we never could have imagined. First and foremost, a garden is never really done. Soil needs long-term care and feeding to get what you need from it. In our case, we started from scratch and had to restore life to our soil. It's hard work, frustrating at times, and as in all dances with nature, rarely dull. In our garden we came to see a microcosm of a could-be world—feed the soil and it will keep feeding us. Not just our bellies, but our minds and spirits too.

We didn't grasp these things at first. After all, we'd set out to renovate our yard, not reframe how we thought about nature. But the process of rebuilding our soil to support a garden revealed the microbial roots of life and health. This new way to read Earth's story redefined, and revived, our relationship with nature—and showed us how to restore the land and heal ourselves.

Along the way, we went from being cynical eco-pessimists to cautious eco-optimists. It didn't take joining a cult, or setting out on a soul-searching global pilgrimage. We opened the back door and stepped outside and began to unearth the wonders around us, beneath us, and within us. New life caught our eye bit by bit and season by season. And when we tapped into our mind's eye, we saw much further, to the invisible frontiers where modern science meets ancient realities.

Most of us think of nature as the plants and animals big enough to see with the naked eye. We too hang on to this tendency. When we look at a tree, we see branches sweeping upward and the shape and color of leaves against the blue of the sky. But in our mind's eye we see so much more that was hidden before. For as unique as we each may be, we have never been alone. The living roots of the grandest tree of all, nature herself, plunge deep beneath our feet and throughout our bodies. Nature is not out there in some distant and faraway land. She is closer than we ever imagined, right inside of us.

Unus pro omnibus, omnes pro uno

One for all, all for one!

Glossary

16S rRNA: Shorthand for 16S ribosomal RNA, a gene common to all life that is needed to make ribosomes, the organelle that assembles proteins.

Adaptive immunity: The type of lifetime immunity found in nearly all vertebrates. Once exposed to specific microbes, adaptive immune cells (B cells and T cells) develop memory and can recognize them thereafter.

Amino acid: An organic molecule used to build larger proteins. There are twenty standard amino acids from which all proteins are made.

Ammonium (NH_4^+): A soluble form of nitrogen (N) that plants can take up; another is nitrate (NO_3^-).

Anaerobic: Occurring in the absence of oxygen; can refer to chemical reactions that proceed without oxygen (like fermentation) or organisms that live in habitats without oxygen (like many archaea).

Antibiotic: Any antimicrobial compound, but generally used in reference to one that kills bacteria.

Antibody: A protein produced by B cells that recognizes and binds to antigen, thereby tagging microbes (or human cells in the case of autoimmunity) made of that antigen for other immune cells to destroy. B cells make antibodies for a wide range of antigen.

Antigen: A molecular sample, typically from a microbe, that helps activate adaptive immune cells (B cells and T cells) to their final developmental stage.

Archaea: Single-celled organisms with key structural differences that make them distinct from bacteria.

Autoimmunity: An immune response against an organism's own cells.

B cell: A type of adaptive immune system cell that specializes in making antibodies.

Bacterial culture: A colony of bacteria grown in a laboratory.

Commensal: In ecology, an organism that lives inside of or on a host without causing harm. In the context of the microbiome, some commensals can become pathogenic when other factors change, such as environmental conditions or the microbial community.

Complex carbohydrates: Long-chain sugar molecules abundant in plant foods.

Cytokine: From *cyto* for "cell" and *kinos* for "movement." Cytokines are signaling molecules with a broad range of functions in immunity generally categorized as having pro- or anti-inflammatory effects.

Dendritic cell: A type of innate immune cell that presents antigen to T cells. Dendritic cells are found primarily where host tissues like the skin, gut, and respiratory tract interface with the external environment. They are instrumental in both pro-inflammatory and anti-inflammatory responses.

Dysbiosis: Imbalance or disruption in the commensal microbial populations of an organism, often associated with ill health.

Endotoxin: A component of the outer membrane of certain bacteria that innate immune cells detect. Too much endotoxin in the circulatory system causes chronic inflammation.

Enzyme: A protein that acts as a catalyst increasing the rate of a reaction. Bacteria in the soil and the human gut make an array of enzymes to break down different kinds of organic matter.

Eukaryotes: The organisms (protists, plants, fungi, and animals) whose cell(s) contain(s) a nucleus that houses their genetic material.

Fermentation: A metabolic pathway used to convert food into forms of usable energy that does not require oxygen. Fermentation of sugars yields acids or alcohols as by-products. Before Earth's oxygen-rich atmosphere existed, many of the earliest life forms, such as archaeans, were fermenters.

Fiber: A general term for the parts of plant foods that people cannot digest. Fiber consists of mostly complex carbohydrates, but also noncarbohydrates (e.g., lignin). Bacteria in the colon ferment the complex carbohydrates in fiber, but the noncarbohydrate portion passes through undigested.

GALT: The immune tissue and cells encircling the digestive tract referred to as gut-associated lymphoid tissue. The majority of the immune system in the human body is GALT.

Genome: The sum total of an organism's genetic material, whether microbe, person, or plant.

Germ theory: The idea that specific microorganisms cause a particular disease, attributed to pioneering nineteenth-century microbiologist Robert Koch.

Germ-free mice: Mice raised to have no microbes living on or within them.

Glacial till: Unsorted glacial sediment, typically a hard-packed mix of clay, sand, gravel, and boulders.

Guano: The excrement of birds or bats, which makes excellent fertilizer rich in nitrogen and phosphorus. Nineteenth-century mining of guano deposits on islands off the coast of South America exhausted commercial supplies.

Horizontal gene transfer: The transfer of genes between organisms without sex.

Host: An organism that is the home of another organism. Typically, the host is the larger of the two organisms. The smaller organism can be commensal, wholly beneficial, or a parasite.

Humus: Dark, carbon-rich organic matter found in soil that is formed from the decay of plant and animal remains that resists further breakdown.

Hyphae: The root-like part of fungi that grow beneath the ground without limit; collectively called mycelium.

Inflammation: From Latin for "ignite." An immune system response to injury, pathogens, or other factors. Inflammation spurs immune-cell activity, changes in blood flow, and causes cytokines to circulate. Inflammation may be either acute or chronic, helpful or harmful.

Innate immunity: The type of immunity found in vertebrates and invertebrates in which particular immune cells can recognize and respond to a wide range of microbes without any prior exposure to the microbe. Dendritic cells and macrophages are types of innate immune cells.

Interleukin: A type of cytokine (signaling molecules secreted by immune cells). Interleukin 6 and interleukin 17 are among cytokines that promote inflammation, whereas interleukin 10 quells inflammation.

Lymph node: Specialized area of immune tissue along a lymph vessel that ranges from millimeters to a centimeter in size. Lymph nodes occur throughout the body and serve as sites where B cells, T cells, and other immune cells congregate, exchange information, and undergo activation.

Macronutrients: Elements and minerals needed in relatively large amounts to build tissues and organs in plants and animals. Carbon and nitrogen are among the macronutrients.

Macrophage: From Greek for "big eater." A type of innate immune cell that presents antigen and engulfs pathogens and cellular debris.

Metabolite: Molecules or compounds that are by-products of an organism's metabolism. The metabolites of many resident microbes in plants and animals are critical for normal growth and development, and in maintaining long-term health of the host.

Microbiome: The sum total of all the genes of the microorganisms that inhabit a host. Also refers to the particular microbiota, or microbial communities, of a host.

Microbiota: The collective microbial inhabitants of an ecosystem or a host.

Micronutrients: In nutrition, the term used for naturally occurring minerals (elements) found in foods that are needed in very small quantities and critical to health in animals and plants. Micronutrients facilitate the action of enzymes. Manganese and zinc are types of micronutrients.

Mineralization: The process by which microbes convert insoluble compounds in soil and organic matter to soluble forms that plants can absorb and use.

Mycorrhizal fungi: Fungi that form symbiotic relationships with plants. The fungi scavenge nutrients from soil or rocks and transport them through their hyphae to plant roots and exchange them for carbohydrates that plants make through photosynthesis.

NPK: Shorthand for nitrogen, phosphorus, and potassium, the big three elements in synthetically produced fertilizers.

Nematode: A diverse group of microscopic worm-like creatures smaller than a grain of sand. Nematodes prey on protists as well as bacteria, and produce micromanure rich in nitrogen.

Nitrate (NO_3^-): A soluble form of nitrogen (N) that plants can take up; another is ammonium (NH_4^+).

Nitrite (NO_2^-): A form of nitrogen (N) that plants cannot take up. Certain soil microbes can convert nitrite to nitrate.

Phosphate (PO_4^{3-}): A soluble form of phosphorus (P) that plants can take up.

Photosynthesis: The process through which plants convert carbon dioxide and water into complex carbohydrates (sugars). Chloroplasts, organelles descended from once-free-living bacteria, are the primary site of photosynthesis.

Phytochemicals: Compounds produced by plants that have a wide range of functions related to defense and health, including communication with microbes.

Polysaccharides: *See* Complex carbohydrates.

Prokaryotes: Single-celled organisms without a cell nucleus. Bacteria and archaea are prokaryotes.

Protists: A very large and diverse group of microbial eukaryotes that compose the kingdom Protista. Algae, amoebas, and slime molds are types of protists. Many protists were formerly called protozoa.

Simple carbohydrates: Short-chain sugar molecules, like glucose, sucrose (table sugar), and fructose (found in fruits). Relative to long-chain sugar molecules, they are readily absorbed in the small intestine.

Symbiogenesis: The origin of new life forms as a result of two (or more) organisms forming a symbiotic relationship. There is strong evidence that different types of free-living microbes physically merging with one another led to multicellular life.

Symbiosis: A close physical and mutually beneficial relationship between two or more organisms belonging to different species.

T cell: A type of adaptive immune cell typically activated by antigen carried on dendritic cells. Among the types of T cells are killer T cells, which kill tumor and infected cells;

regulatory T cells (Tregs), which quell inflammation; and Th17 cells, which promote inflammation.

Vaccine: A harmless form of a disease-causing microbe that stimulates the adaptive immune system to respond to the microbe upon subsequent exposure, thereby conferring immunity. Vaccines are typically made from weakened or killed microbes, or from their distinctive surface proteins.

Notes

INTRODUCTION

1. Maynard et al., 2012, 233.

CHAPTER 1: DEAD DIRT

1. The scientific community debates the benefits of soil soup and other types of compost tea. Of course, gardens vary like any natural system. Soils, plants, insects, microclimates, and other factors change from place to place. Plus, everyone applies compost tea differently and at different times of the day and year.

CHAPTER 2: THINKING SMALL

1. It sounds strange to us that fungi in general are considered microbes. But there is a reason. Fungi are unicellular. Their cells are not completely separate and closed off from one another. At the scale of a fungal cell, fairly sizable canals exist where two cells touch, allowing fluids, proteins, and even nuclei to freely move between cells.
2. Although it is not entirely accurate, we will hereafter follow common usage of the term "antibiotic" to indicate drugs that kill bacteria. Strictly speaking, however, drugs that kill bacteria are antibacterials, and those that kill fungi are antifungals. Likewise, drugs that thwart viruses are called antivirals.
3. Dog lovers may be interested to know that the rumen is the first thing that wolves devour after taking down an herbivore. They may be after the vitamin B12 that methanogenic archaea in the rumen produce during anaerobic fermentation. Or maybe wolves just love the stench. While people find the smell of rumen overpowering, dogs love it. Savvy pet food manufacturers add rumen bacteria to make kibble smell more appealing to the canine palate.

CHAPTER 3: LOOKING INTO LIFE

1. The double helix structure of DNA is like a ladder twisted round and round. Pairs of four base molecules—adenine, thymine, cytosine, and guanine—form the "rungs" of the ladder. Long sequences of the four base molecules running lengthwise along the "ladder" make up different genes.

2. The "genetic code" refers to how particular threesomes of adjacent base molecules in DNA specify amino acids, which in turn are linked together to make proteins.
3. As we were in the final stages of editing, researchers from Uppsala University reported finding what media reports characterized as the missing link between prokaryotes and eukaryotes. In deep marine sediments from the Arctic Mid-Ocean Ridge, the scientists found genetic evidence of archaea that shared eukaryotic signature traits. They propose that these new archaea, the Lokiarchaeota, are a living example of the kind of ancient archaea that gave rise to the first eukaryotes.
4. It's estimated that only about one-quarter of the more than 100 phyla of bacteria known today are amenable to easy culturing.

CHAPTER 4: BETTER TOGETHER

1. Margulis, 1998, 26–27.
2. Margulis, 1998, 37.
3. Strictly speaking, however, some protists, like giardia, have lost their mitochondria but retain relict structures known as mitosomes.
4. In 2014, as we were writing this book, two Harvard microbiologists conducted an experiment in which they established a symbiotic relationship between a green algae (*Chlamydomonas reinhardtii*) and a yeast (*Saccharomyces cerevisiae*, a favorite of brewers since ancient times). The yeast metabolized glucose, producing carbon dioxide that the alga used in photosynthesis. Conversely, the alga metabolized nitrite, producing ammonia, which the yeast could use as a nitrogen source.

CHAPTER 5: WAR ON THE SOIL

1. In all but the driest regions, covering an area with cardboard is a way to eliminate plants (especially grass and dandelion infestations) without using poison or digging. This method works very well unless the plants you are trying to eliminate have a particularly hearty root system (e.g., morning glory) or are just plain ornery (e.g., horsetail or ivy). If you want to do this in your garden, here are a few tips:
   (a) Use cardboard that has no paint or ink (or run a razor blade around the offending part and then pull it off) and remove all glue, stickers, labels, tape, and staples.
   (b) Place the cardboard methodically and overlap it where two pieces meet.
   (c) Layer no less than 3 or 4 inches of wood chips on top of the cardboard. Let the cardboard sit, preferably over a growing season, and then push aside the wood chips and cut through the cardboard to plant.
2. Howard, 1940, 161.
3. Howard, 1940, 189.
4. Howard, 1940, 220.
5. Howard, 1940, 51.
6. Howard, 1940, 61, 166.
7. Howard, 1946, 57.

CHAPTER 6: UNDERGROUND ALLIES

1. Different isotopes of an element have the same number of protons but different numbers of neutrons, and thus a different atomic mass. Carbon-13 (^{13}C) is a rela-

tively rare stable isotope of carbon that can be used as a natural tracer when introduced in higher than background concentrations to living organisms.

2. Other species of termites rely on symbiotic bacteria in their gut. These termites may bite off chunks of wood, but they can't digest them. The microbes in their gut do that for them. Scientists figured this out by killing off the gut flora of termites with antibiotics and watching what happened. The termites went about their wood-eating business, but within days they started to die. Without their gut microbes, they couldn't get any metabolic energy from the cellulose in the wood. They ate their fill and starved to death.

CHAPTER 7: TOO CLOSE TO HOME

1. In 2013, the American Academy of Microbiology (ASM) reviewed basic information about the ratio of microbial cells to human cells. They revised the often-cited figure of 10 bacterial cells for each human cell, found on the National Institutes of Health website, to 3 bacterial cells for each human cell. The ASM also reported that our microbiome could have as many as 5 viruses for each bacterium, and that bacteria outnumber fungi by about 10 to 1.

CHAPTER 8: INNER NATURE

1. For more extended discussions of the changing incidence of common autoimmune diseases, see Velasquez-Manoff (2012) and Blaser (2014).
2. "Cytokine" comes from *cyto* for "cell" and *kinos* for "movement." These molecules have a broad range of functions, often initiating or inhibiting the actions of immune cells. For example, one type of cytokine, called a chemokine, attracts immune cells to the site of a wound or infection.
3. Although "commensal" is the commonly used term in microbiome research, it is not quite accurate. In ecology a commensal relationship is one in which one member benefits and the other member is not affected positively or negatively. "Situational symbiont" might better describe the hat changers, but no such term officially exists.
4. To date, the regular and widespread presence of segmented filamentous bacteria in people remains a mixed bag. Researchers found them in abundance in infants up to three years old and in lower numbers thereafter in the lower part of the small intestine.

CHAPTER 9: INVISIBLE ENEMIES

1. Stearns, 1950, 115.

CHAPTER 10: FEUDING SAVIORS

1. Hopefully, for these pathogens and others, there may be a way to blunt their counterattack. As we finished writing this book, a study published online in the journal *Nature* reported the discovery of teixobactin, a new antibiotic derived from previously unculturable soil bacteria. The researchers used a novel method to grow colonies of the teixobactin producer in a growth chamber that mimics the bac-

terium's natural soil environment. Teixobactin interferes with the production of lipid molecules that a large group of bacteria make to build their cell walls. These molecules are so fundamental to bacterial integrity that the study's authors think it will take decades, or more, for bacteria to develop resistance to teixobactin. Among the pathogens that the new antibiotic effectively killed are those that cause tuberculosis, methicillin-resistant *Staphylococcus aureus* (MRSA), anthrax, and *C. difficile.* This exciting discovery reveals the power of soil bacteria in nature's pharmacy. Should new antibiotics like teixobactin make it through to commercial production, however, what will this mean for the future of our microbiome? That, of course, will depend on how we use them.

CHAPTER 11: PERSONAL ALCHEMISTS

1. Aside from the obvious differences between mice and people, one also has to wonder about the nature of mouse chow itself. We did. For example, mice are herbivores. So, are the fats in mouse chow from the types of seeds a wild mouse would eat, or from animal fat? And if from an animal, say a cow, what did the cows eat, grains or grass? We'll leave these questions for others to explore.
2. It's important to note that the obese mice only became lean, when, in addition to eating their lean cage mate's feces, they also ate a low-fat, high-fiber diet. When the obese mice ate a high-fat, low-fiber diet, the lean microbiota could not successfully colonize the mice, and the obese mice remained so.

CHAPTER 12: TENDING THE GARDEN

1. "Dietary fiber" is a confusing term, with several different meanings. It generally refers to the indigestible parts of plant foods and consists of both carbohydrates and noncarbohydrate molecules. Our gut microbiota handily ferment the carbohydrate portion of fiber. But neither they nor we have the enzymes to crack the noncarbohydrate portion of fiber. Lignin, a component of wood, is one example of the noncarbohydrate fraction of fiber. Not being made of sugars, ligin offers nothing for our gut microbiota to ferment. No doubt some of the confusion about the term arises because the carbohydrate and noncarbohydrate fractions are both found in whole plant foods. When we use "fiber," we mean the fraction made of complex carbohydrates from plants that our gut microbiota ferment that yield SCFAs. However, if your doctor or spouse nags you to eat more fiber, it is likely for a different reason. The noncarbohydrate portion provides the heft and bulk necessary for a stool to form, and most important, keep moving.
2. Like us, you may have no idea how much fiber you actually eat. For reference, a medium-size apple contains about 4 grams of fiber and half a cup of black beans contains almost 8 grams of fiber.
3. Edible fermentable polysaccharides are not exclusive to the plant kingdom. Other sources include seaweeds, algae, and fungi, and even some types of animal tissues. If you are among the legions of people who lack the enzyme lactase, your gut microbiota will ferment the milk sugars in cheeses, milk, and other dairy products (which may leave you feeling unwell). But if you have this enzyme, you break down dairy products yourself and nothing is left for the gut microbiota. Meat possesses a tiny fraction of fermentable carbohydrates since sugars are stored in the muscle tissue of animals. And in another case, human breast milk provides a rich source

of fermentable sugars that help establish and keep the infant gut microbiome well nourished.

4. As far as getting probiotics into the vagina, that's pretty straightforward. Just insert a capsule or gel containing the desired dosage. Or you can send them through the gut to the bottom of the digestive tract. From there it's a short jaunt out of the rectum, along the perineum, and up the vagina. Researchers have tested both routes and found them equally effective in delivering lactobacilli to where they are needed.
5. Far more controversial than the effectiveness of fecal transplants is how to regulate them. While the Food and Drug Administration has opted to treat fecal matter as a drug, a classification that typically requires years of testing and clinical trials, researchers are calling for treating fecal transplants as human tissue, which would allow for much faster and wider medically supervised adoption. Fueling this debate are concerns over the safety and growing popularity of home fecal transplants.
6. The average American consumes under a liter of olive oil a year. That bottle in your cupboard is most likely a half-liter. To eat like a Cretan you'd need to down more than a bottle a week.

CHAPTER 13: COURTING ANCIENT FRIENDS

1. Cotton gin trash is the material left over from the processing of cotton, consisting of chaff, seeds, and other residual plant material.
2. If you don't eat enough fermentable carbohydrates some gut bacteria can turn against you, overgrazing the mucosal lining of the gut, which can cause serious problems.

Sources

INTRODUCTION

Maynard, C. L.; Elson, C. O.; Hatton, R. D.; and Weaver, C. T. 2012. Reciprocal interactions of the intestinal microbiota and immune system, *Nature*, v. 489, pp. 231–241.

CHAPTER 2: THINKING SMALL

Ben-Barak, I. 2009. *The Invisible Kingdom: From the Tips of Our Fingers, to the Tops of Our Trash, Inside the Curious World of Microbes*. New York: Basic Books, 204 pp.

Bonneville, S., et al. 2009. Plant-driven fungal weathering: Early stages of mineral alteration at the nanometer scale. *Geology*, v. 37, pp. 615–618.

Brodie, E. L., et al. 2007. Urban aerosols harbor diverse and dynamic bacterial populations. *Proceedings of the National Academy of Sciences*, v. 104, pp. 299–304.

Burrows, S. M.; Elbert, W.; Lawrence, M. G.; and Pöschl, U. 2009. Bacteria in the global atmosphere. Part 1: Review and synthesis of literature data for different ecosystems. *Atmospheric Chemistry and Physics*, v. 9, pp. 9263–9280.

Christner, B. C., et al. 2014. A microbial ecosystem beneath the West Antarctic ice sheet. *Nature*, v. 512, pp. 310–313.

Fahlgren, C.; Hagström, A.; Nilsson, D.; and Zweifel, U. L. 2010. Annual variations in the diversity, viability, and origin of airborne bacteria. *Applied and Environmental Microbiology*, v. 76, pp. 3015–3025.

Fierer, N., et al. 2012. Cross-biome metagenomic analyses of soil microbial communities and their functional attributes. *Proceedings of the National Academy of Sciences*, v. 109, pp. 21,390–21,395.

Gazzè, S. A., et al. 2012. Nanoscale channels on ectomycorrhizal-colonized chlorite: Evidence for plant-driven fungal dissolution. *Journal of Geophysical Research: Biogeosciences*, v. 117, p. G00N09, doi:10.1029/2013JG002016.

Holloway, J. M., and Dahlgren, R. A. 2002. Nitrogen in rock: Occurrences and biogeochemical implications. *Global Biogeochemical Cycles*, v. 16, p. 1118, doi:10.1029/2002GB001862.

Ingraham, J. L. 2010. *March of the Microbes: Sighting the Unseen*. Cambridge and London: Belknap Press of Harvard University Press, 326 pp.

Khelaifia, S., and Drancourt, M. 2012. Susceptibility of archaea to antimicrobial agents: Applications to clinical microbiology. *Clinical Microbiology and Infection*, v. 18, pp. 841–848.

Kolter, R., and Maloy, S., eds. 2012. *Microbes and Evolution: The World That Darwin Never Saw*. Washington, D.C.: ASM Press, 299 pp.

Lanter, B. B.; Sauer, K.; and Davies, D. G. 2014. Bacteria present in carotid arterial plaques are found as biofilm deposits which may contribute to enhanced risk of plaque rupture. *mBio*, v. 3; pp. e01206–14.

Lepot, K.; Benzerara, K.; Brown, G. E.; and Philippot, P. 2008. Microbially influenced formation of 2,724-million-year-old stromatolites. *Nature Geoscience*, v. 1, pp. 118–121.

Lyons, T. W.; Reinhard, C. T.; and Planavsky, N. J. 2014. The rise of oxygen in Earth's early ocean and atmosphere. *Nature*, v. 506, pp. 307–315.

Mattes, T. E., et al. 2013. Sulfur oxidizers dominate carbon fixation at a biogeochemical hot spot in the dark ocean. *ISME Journal*, v. 7, pp. 2349–2360.

McCarthy, M. D., et al. 2011. Chemosynthetic origin of ^{14}C-depleted dissolved organic matter in a ridge-flank hydrothermal system. *Nature Geoscience*, v. 4, pp. 32–36.

Orsi, W. D.; Edgcomb, V. P.; Christman, G. D.; and Biddle, J. F. 2013. Gene expression in the deep biosphere. *Nature*, v. 499, pp. 205–208.

Overballe-Petersen, S., et al. 2013. Bacterial natural transformation by highly fragmented and damaged DNA. *Proceedings of the National Academy of Sciences*, v. 110, pp. 19,860–19,865.

Planavsky, N. J., et al. 2014. Low Mid-Proterozoic atmospheric oxygen levels and the delayed rise of animals. *Science*, v. 346, pp. 635–638.

Reyes, L., et al. 2013. Periodontal bacterial invasion and infection: Contribution to atherosclerotic pathology. *Journal of Periodontology*, v. 84 (4 Suppl.), pp. S30–S50.

Sattler, B.; Puxbaum, H.; and Psenner, R. 2001. Bacterial growth in supercooled cloud droplets. *Geophysical Research Letters*, v. 28, pp. 239–242.

Schönknecht, G., et al. 2013. Gene transfer from bacteria and archaea-facilitated evolution of an extremophilic eukaryote. *Science*, v. 339, pp. 1207–1210.

Smith, D. J. 2011. Microbial survival in the stratosphere and implications for global dispersal. *Aerobiologia*, v. 27, pp. 319–332.

Smith, D. J., et al. 2013. Intercontinental dispersal of bacteria and archaea in transpacific winds. *Applied and Environmental Microbiology*, v. 79, pp. 1134–1139.

Wacey, D., et al. 2011. Microfossils of sulphur-metabolizing cells in 3.4-billion-year-old rocks of Western Australia. *Nature Geoscience*, v. 4, pp. 698–702.

Walter, M. R.; Buick, R.; and Dunlop, J. S. R. Stromatolites 3,400–3,500M yr old from the North Pole area, Western Australia. *Nature*, v. 284, pp. 443–445.

Wecht, K. J., et al. 2014. Mapping of North American methane emissions with high spatial resolution by inversion of SCIAMACHY satellite data. *Journal of Geophysical Research: Atmospheres*, v. 119, pp. 7741–7756.

Whitman, W. B.; Coleman, D. C.; and Wiebe, W. J. 1998. Prokaryotes: The unseen majority. *Proceedings of the National Academy of Sciences*, v. 95, pp. 6578–6583.

CHAPTER 3: LOOKING INTO LIFE

De Kruif, P. 1926. *Microbe Hunters*. New York: Harcourt, Brace & Co., 363 pp.

Dobel, C., 1958. *Antony van Leeuwenhoek and His "Little Animals" Being Some Account of the Father of Protozoology & Bacteriology and His Multifarious Discoveries in These Disciplines*. New York: Russell & Russell, 435 pp.

Ford, B. J. 1991. *The Leeuwenhoek Legacy*. Bristol and London: Biopress and Farrand Press, 185 pp.

Gilbert, J. A.; van der Lelie, D.; and Zarraonaindia, I. 2014. Microbial *terroir* for wine grapes. *Proceedings of the National Academy of Sciences*, v. 111, pp. 5–6.

Gold, L. 2013. The kingdoms of Carl Woese. *Proceedings of the National Academy of Sciences*, v. 110, pp. 3206–3207.

Gould, S. J. 2002. *The Structure of Evolutionary Theory*. Cambridge and London: Belknap Press of Harvard University Press, 1433 pp.

Ingraham, J. L. 2010. *March of the Microbes: Sighting the Unseen*. Cambridge and London: Harvard University Press, 326 pp.

Kolter, R., and Maloy, S., eds. 2012. *Microbes and Evolution: The World That Darwin Never Saw*. Washington, D.C.: ASM Press, 299 pp.

Mojzsis, S. J., et al. 1996. Evidence for life on Earth before 3,800 million years ago. *Nature*, v. 384, pp. 55–59.

Nair, P. 2012. Woese and Fox: Life, rearranged. *Proceedings of the National Academy of Sciences*, v. 109, pp. 1019–1021.

Pace, N. R.; Sapp, J.; and Goldenfeld, N. 2012. Phyology and beyond: Scientific, historical, and conceptual significance of the first tree of life. *Proceedings of the National Academy of Sciences*, v. 109, pp. 1011–1018.

Spang, A., et al. 2015. Complex archaea that bridge the gap between prokaryotes and eukaryotes. *Nature*, v. 521, pp. 173–179.

Woese, C. R. 2004. A new biology for a new century. *Microbiology and Molecular Biology Reviews*, v. 68, pp. 173–186.

Woese, C. R., and Fox, G. E. 1977. Phylogenetic structure of the prokaryotic domain: The primary kingdoms. *Proceedings of the National Academy of Sciences*, v. 74, pp. 5088–5090.

Woese, C. R.; Kandler, O.; and Wheelis, M. L. 1990. Towards a natural system of organisms: Proposal for the domains Archaea, Bacteria, and Eucarya. *Proceedings of the National Academy of Sciences*, v. 87, pp. 4576–4579.

CHAPTER 4: BETTER TOGETHER

Archibald, J. 2014. *One Plus One Equals One: Symbiosis and the Evolution of Complex Life*. Oxford: Oxford University Press, 205 pp.

Brock, D. A.; Douglas, T. E.; Queller, D. C.; and Strassmann, J. E. 2011. Primitive agriculture in a social amoeba. *Nature*, v. 469, pp. 393–396.

Chapela, I. H.; Rehner, S. A.; Schulta, T. R.; and Meuller, U. G. 1994. Evolutionary history of the symbiosis between fungus-growing ants and their fungi. *Science*, v. 266, pp. 1691–1694.

Dolan, M. F., and Margulis., L. 2011. *Hans Ris 1914–2002, A Biographical Memoir*. Washington, D.C.: National Academy of Sciences, 16 p.

Domazet-Loso, T., and Tautz, D. 2008. An ancient evolutionary origin of genes associated with human genetic diseases. *Molecular Biology and Evolution*, v. 25, pp. 2699–2707.

Farrell, B. D., et al. 2001. The evolution of agriculture in beetles (Curculionidae: Scolytinae and Platypodinae). *Evolution*, v. 55, pp. 2011–2027.

Hom, E. F. Y., and Murray, A. W. 2014. Niche engineering demonstrates a latent capacity for fungal-algal mutualism. *Science*, v. 345, pp. 94–98.

Kozo-Polyansky, B. M. 2010 (1924). *Symbiogenesis: A New Principle of Evolution*. Trans. Fet, V. Ed. Fet, V., and Margulis, L. Cambridge and London: Harvard University Press, 198 pp.

Margulis (Sagan), L. 1967. On the origin of mitosing cells. *Journal of Theoretical Biology*, v. 14, pp. 225–274.

Margulis, L., 1998. *Symbiotic Planet*. New York: Basic Books, 147 pp.

Margulis, L., and Sagan, D. 1986. *Microcosmos: Four Billion Years of Evolution from Our Microbial Ancestors*. New York: Summit Books, 301 pp.

Mueller, U. G., et al. 2001. The origin of the attine ant-fungus mutualism. *Quarterly Review of Biology*, v. 76, pp. 169–197.

O'Connor, R. M., et al. 2014. Gill bacteria enable a novel digestive strategy in a wood-feeding mollusk. *Proceedings of the National Academy of Sciences*, v. 111, pp. E5096–E5104.

Pennisi, E. 2014. Modern symbionts inside cells mimic organelle evolution. *Science*, v. 346, pp. 532–533.

Scott, J. J., et al. 2008. Bacterial protection of beetle-fungus mutualism. *Science*, v. 322, p. 63.

Shih, P. M., and Matzke, N. J. 2013. Primary endosymbiosis events date to later Proterozoic with cross-calibrated dating of duplicated ATPase proteins. *Proceedings of the National Academy of Sciences*, v. 110, pp. 12,355–12,360.

Yoon, C. K. 2009. *Naming Nature: The Clash Between Instinct and Science.* New York and London: W. W. Norton, 341pp.

CHAPTER 5: WAR ON THE SOIL

Behie, S. W.; Zilisco, P. M.; and Bidochka, M. J. 2012. Endophytic insect-parasitic fungi translocate nitrogen directly from insects to plants. *Science*, v. 336, pp. 1576–1577.

Heckman, J. 2006. A history of organic farming: Transitions from Sir Albert Howard's *War in the Soil* to USDA National Organic Program. *Renewable Agriculture and Food Systems*, v. 21, pp. 143–150.

Hershey, D. 2003. Misconceptions about Helmont's willow experiment. *Plant Science Bulletin*, v. 49, pp. 78–84.

Howard, A., 1940 (1945), *An Agricultural Testament.* London, New York, and Toronto: Oxford University Press, 253 p.

Howard, A. 1946. *The War in the Soil.* Emmaus, Pa.: Rodale Press, 96 pp.

Johnston, A. E., and Mattingly, G. E. G. 1976. Experiments on the continuous growth of arable crops at Rothamsted and Woburn experimental stations: Effects of treatments on crop yields and soil analyses and recent modifications in purpose and design. *Annals of Agronomy*, v. 27, pp. 927–956.

Kassinger, R. 2014. *A Garden of Marvels: How We Discovered That Flowers Have Sex, Leaves Eat Air, and Other Secrets of Plants.* New York: William Morrow, 416 pp.

Montgomery, D. R. 2007. *Dirt: The Erosion of Civilizations.* Berkeley: University of California Press, 285 pp.

Mortford, S.; Houlton, B. Z.; and Dahlgren, R. A. 2011. Increased forest nitrogen and carbon storage from nitrogen-rich bedrock. *Nature*, v. 477, pp. 78–81.

Pagel, W. 1982. *Joan Baptista Van Helmont: Reformer of Science and Medicine.* Cambridge: Cambridge University Press, 232 pp.

Shenstone, W. A. 1895. *Justus von Liebig: His Life and Work (1803–1873).* London, Paris, and Melbourne: Cassell & Co., 219 pp.

CHAPTER 6: UNDERGROUND ALLIES

Akiyama, K.; Matsuzaki, K.; Hayashi, H. 2005. Plant sesquiterpenes induce hyphal branching in arbuscular mycorrhizal fungi. *Nature*, v. 435, pp. 824–827.

Bais, H. P., et al. 2006. The role of root exudates in rhizosphere interactions with plants and other organisms. *Annual Review of Plant Biology*, v. 57, pp. 233–266.

Behrensmeyer, A. K., et al. 1992. *Terrestrial Ecosystems Through Time: Evolutionary Paleoecology of Terrestrial Plants and Animals.* Chicago and London: University of Chicago Press, 568 pp.

Berendsen, R. L.; Pieterse, C. M. J.; and Bakker, P. A. H. M. 2012. The rhizosphere microbiome and plant health. *Trends in Plant Science*, v. 17, pp. 478–486.

Berg, G., and Smalla, K. 2009. Plant species and soil type cooperatively shape the structure and function of microbial communities in the rhizosphere. *FEMS Microbiology Ecology*, v. 68, pp. 1–13.

Bonkowski, M.; Villenave, C.; and Griffiths, B. 2009. Rhizosphere fauna: The functional and structural diversity of intimate interactions of soil fauna with plant roots. *Plant and Soil*, v. 321, pp. 213–233.

Boyce, C. K., et al. 2007. Devonian landscape heterogeneity recorded by a giant fungus. *Geology*, v. 35, pp. 399–402.

Brigham, L. A.; Michaels, P. J.; and Flores, H. E. 1999. Cell-specific production and antimicrobial activity of napthoquinones in roots of *Lithospermum erythrorhizon*. *Plant Physiology*, v. 119, pp. 417–428.

Broeckling, C. D., et al. 2008. Root exudates regulate soil fungal community composition and diversity. *Applied Environmental Microbiology*, v. 74, pp. 738–744.

Bulgarelli, D., et al. 2013. Structure and functions of the bacterial microbiota of plants. *Annual Review of Plant Biology*, v. 64, pp. 807–838.

Cesco, S., et al. 2010. Release of plant-borne flavonoids into the rhizosphere and their role in plant nutrition. *Plant and Soil*, v. 329, pp. 1–25.

Christensen, M. 1989. A view of fungal ecology. *Mycologia*, v. 81, pp. 1–19.

Clark, F. E. 1949. Soil microorganisms and plant roots. *Advances in Agronomy*, v. 1, pp. 241–288.

Dakora, F. D., and Phillips, D. A. 2002. Root exudates as mediators of mineral acquisition in low-nutrient environments. *Plant and Soil*, v. 245, pp. 35–47.

Dennis, P. G.; Miller, A. J.; and Hirsch, P. R. 2010. Are root exudates more important than other sources of rhizodeposits in structuring rhizosphere bacterial communities? *FEMS Microbiology Ecology*, v. 72, pp. 313–327.

Doty, S. L., et al. 2009. Diazotrophic endophytes of native black cottonwood and willow. *Symbiosis*, v. 47, pp. 23–33.

Dyall, S. D.; Brown, M. T.; and Johnson, P. J. 2004. Ancient invasions: From endosymbionts to organelles. *Science*, v. 304, pp. 253–257.

Farrar, J.; Hawes, M.; Jones, D.; and Lindow, S. 2003. How roots control the flux of carbon to the rhizosphere. *Ecology*, v. 84, pp. 827–837.

Foster, R. C. 1986. The ultrastructure of the rhizoplane and rhizosphere. *Annual Review of Phytopathology*, v. 24, pp. 211–234.

Gaiero, J. R., et al. 2013. Inside the root microbiome: Bacterial root endophytes and plant growth promotion. *American Journal of Botany*, v. 100, pp. 1,738–1,750.

Garcia-Garrido, J. M., and Ocampo, J. A. 2002. Regulation of the plant defense response in arbuscular mycorrihzal symbiosis. *Journal of Experimental Botany*, v. 53, pp. 1377–1386.

Haichar, F. Z., et al. 2008. Plant host habitat and root exudates shape soil bacterial community structure. *ISME Journal*, v. 2, pp. 1221–1230.

Hardoim, P. R.; van Overbeek, L. S.; and van Elsas, J. D. 2008. Properties of bacterial endophytes and their proposed role in plant growth. *Trends in Microbiology*, v. 16, pp. 463–471.

Hartann, A.; Rothballer, M.; and Schmid, M. 2008. Lorenz Hiltner, a pioneer in rhizosphere microbial ecology and soil bacteriology research. *Plant and Soil*, v. 312, pp. 7–14.

Hassan, S., and Mathesius, U. 2012. The role of flavonoids in root-rhizosphere signaling: Opportunities and challenges for improving plant-microbe interactions. *Journal of Experimental Botany*, v. 63, pp. 3429–3444.

Heckman, D. S., et al. 2001. Molecular evidence for the early colonization of land by fungi and plants. *Science*, v. 293, pp. 1129–1133.

Hinsinger, P. 1998. How do plant roots acquire mineral nutrients?: Chemical processes involved in the rhizosphere. *Advances in Agronomy*, v. 64, pp. 225–265.

Hochuli, P. A., and Feist-Burkhardt, S. 2013. Angiosperm-like possen and *Afropollis* from the Middle Triassic (Anisian) of the Germanic Basin (Northern Switzerland). *Frontiers in Plant Science*, v. 4, p. 344, doi:10.3389/fpls.2013.00344.

Horodyski, R. J., and Knauth, L. P. 1994. Life on land in the Precambrian. *Science*, v. 263, pp. 494–498.

Ingraham, J. L. 2010. *March of the Microbes: Sighting the Unseen*. Cambridge and London: Harvard University Press, 326 pp.

Jimenez-Salgado, T., et al. 1997. *Coffea Arabica* L., a new host plant for *Acetobacter diazotrophicus*, and isolation of other nitrogen-fixing Acetobacteria. *Applied and Environmental Microbiology*, v. 63, pp. 3676–3683.

Johnson, J. F.; Allan, D. L.; Vance, C. P.; and Weiblen, G. 1996. Root carbon dioxide fixation by phosphorus-deficient *Lupinus albus*—contribution to organic acid exudation by proteoid roots. *Plant Physiology*, v. 111, pp. 19–30.

Jones, D. L., and Darrah, P. R. 1995. Influx and efflux of organic-acids across the soil-root interface of *Zea mays* L. and its implications in rhizosphere C flow. *Plant and Soil*, v. 173, pp. 103–109.

López-Guerrero, M. G., et al. 2013. Buffet hypothesis for microbial nutrition at the rhizosphere. *Frontiers in Plant Science*, v. 4, pp. 1–4.

Maillet, F., et al. 2011. Fungal lipochitooligosaccharide symbiotic signals in arbuscular mycorrhiza. *Nature*, v. 469, pp. 58–64.

Makoi, J. H. Jr., and Ndakidemia, P. A. 2007. Biological, ecological, and agronomic significance of plant phenolic compounds in rhizosphere of the symbiotic legumes. *African Journal of Biotechnology*, v. 6, pp. 1358–1368.

Martin, F., et al. 2001. Developmental cross talking in the ectomycorrhizal symbiosis: Signals and communication genes. *New Phytologist*, v. 151, pp. 145–154.

Marx, J. 2004. The roots of plant-microbe collaborations. *Science*, v. 304, pp. 234–236.

Masaoka, Y., et al. 1993. Dissolution of ferric phosphates by alfalfa (*Medicago sativa* L.) root exudates. *Plant and Soil*, v. 155, pp. 75–78.

Miltner, A.; Bomback, P.; Schmidt-Brücken, B; and Kästner, M. 2012. SOM genesis: Microbial biomass as a significant source. *Biochemistry*, v. 111, pp. 41–55.

Newman, E. I. 1985. The rhizosphere: carbon sources and microbial populations. In *Ecological Interactions in Soil*. Ed. A. H. Fitter. Oxford: Blackwell Scientific Publications, pp. 107–121.

Perin, L., et al. 2006. Diazotrophic *Burkholderia* species associated with field-grown maize and sugarcane. *Applied and Environmental Microbiology*, v. 72, pp. 3103–3110.

Pühler, A., et al. 2004. What can bacterial genome research teach us about bacteria-plant interactions? *Current Opinion in Plant Biology*, v. 7, pp. 137–147.

Retallack, G. J. 1985. Fossil soils as grounds for interpreting the advent of large plants and animals on land. *Philosophical Transactions of the Royal Society of London*, v. B 309, pp. 108–142.

Retallack, G. J., and Feakes, C. R. 1987. Trace fossil evidence for Late Ordovician animals on land. *Science*, v. 235, pp. 61–63.

Rillig M. C., and Mummey, D. L. 2006. Mycorrhizas and soil structure. *New Phytologist*, v. 171, pp. 41–53.

Rodriguez, H., and Fraga, R. 1999. Phosphate solubilizing bacteria and their role in plant growth promotion. *Biotechnology Advances*, v. 17, pp. 319–339.

Rudrappa, T.; Biedrzycki, M. L.; and Bais, H. P. 2008. Causes and consequences of plant-associated biofilms. *FEMS Microbiology Ecology*, v. 64, pp. 153–166.

Rudrappa, T.; Czymmek, K.; Paré, P. W.; and Bais, H. P. 2008. Root-secreted malic acid recruits beneficial soil bacteria. *Plant Physiology*, v. 148, pp. 1547–1556.

Schurig, C., et al. 2013. Microbial cell-envelope fragments and the formation of soil organic matter: A case study from a glacier forefield. *Biogeochemistry*, v. 113, pp. 595–612.

Turner, T. R.; James, E. K.; and Poole, P. S. 2013. The plant microbiome. *Genome Biology*, v. 14, p. 209.
Vacheron, J., et al. 2013. Plant growth-promoting rhizobacteria and root system functioning. *Frontiers in Plant Science*, v. 4, doi:10.3389/fpls.2013.00356.

CHAPTER 7: TOO CLOSE TO HOME

Akagi, K., et al. 2014. Genome-wide analysis of HPV integration in human cancers reveals recurrent, focal genomic instability. *Genome Research*, v. 24, pp. 185–199.
American Academy of Microbiology. 2014. *Human Microbiome FAQ*. Washington, D.C.: American Society for Microbiology, 16 pp.
Bakhtiar, S. M., et al. 2013. Implications of the human microbiome in inflammatory bowel diseases. *FEMS Microbiology Letters*, v. 342, pp. 10–17.
Balter, M. 2012. Taking stock of the human microbiome and disease. *Science*, v. 336, pp. 1246–1247.
Bianconi, E., et al. 2013. An estimation of the number of cells in the human body. *Annals of Human Biology*, v. 40, pp. 463–471.
Chaturvedi, A. K., et al. 2011. Human Papillomavirus and rising oropharyngeal cancer incidence in the United States. *Journal of Clinical Oncology*, v. 29, pp. 4294–4301.
Chaturvedi, A. K., et al. 2013. Worldwide trends in incidence rates for oral cavity and oropharyngeal cancers. *Journal of Clinical Oncology*, v. 31, pp. 4550–4559.
Costello, E. K., et al. 2012. The application of ecological theory toward an understanding of the human microbiome. *Science*, v. 336, pp. 1255–1262.
Ezkurdia, I., et al. 2014. Multiple evidence strands suggest that there may be as few as 19,000 human protein-coding genes. *Human Molecular Genetics*, v. 23, pp. 5866–5878.
Gordon, J. I. 2012. Honor thy gut symbionts redux. *Science*, v. 336, pp. 1251–1253.
Haiser, H. J., and Turnbaugh, P. J. 2012. Is it time for a metagenomic basis of therapeutics? *Science*, v. 336, pp. 1253–1255.
Hooper, L. V.; Littman, D. R.; and Macpherson A. J. 2012. Interactions between the microbiota and the immune system. *Science*, v. 336, pp. 1268–1273.
International Human Genome Sequencing Consortium. 2004. Finishing the euchromatic sequence of the human genome. *Nature*, v. 431, pp. 931–945.
Lozupone, C. A., et al. 2012. Diversity, stability, and resilence of the human gut microbiota. *Nature*, v. 489, pp. 220–230.
Maynard, C. L.; Elson, C. O.; Hatton, R. D.; and Weaver, C. T. 2012. Reciprocal interactions of the intestinal microbiota and immune system. *Nature*, v. 489, pp. 231–241.
Mesri, E. A., Feitelson, M. A., Munger, K. 2014. Human viral oncogenesis: A cancer hallmarks analysis. *Cell Host & Microbe*, v. 15, pp. 266–282.
Qin, J., et al. 2010. A human gut microbial gene catalogue established by metagenomic sequencing. *Nature*, v. 464, pp. 59–65.
Ramqvist, T., and Dalianis, T. 2010. Oropharyngeal cancer epidemic and human papillomavirus. *Emerging Infectious Diseases*, v. 16, pp. 1671–1677.
Relman, D. A. 2012. Learning about who we are. *Nature*, v. 486, pp. 194–195.
Servan-Schreiber, D. 2009. *Anticancer: A New Way of Life*. New York: Viking, 272 pp.
The Human Microbiome Project Consortium. 2012. A framework for human microbiome research. *Nature*, v. 486, pp. 215–221.
The Human Microbiome Project Consortium. 2012. Structure, function, and diversity of the healthy human microbiome. *Nature*, v. 486, pp. 207–214.
Tremaroli, V., and Bäckhed, F. 2012. Functional interactions between the gut microbiota and host metabolism. *Nature*, v. 489, pp. 242–249.

Vidal, A. C., et al. 2014. HPV genotypes and cervical intraepithelial neoplasia in a multiethnic cohort in the southeastern United States. *Journal of Vaccines and Vaccination*, v. 5, p. 224, doi:10.4172/2157-7560.1000224.

CHAPTER 8: INNER NATURE

Atarashi, K., et al. 2011. Induction of colonic regulatory T cells by indigenous *Clostridium* species. *Science*, v. 331, pp. 337–341.

Atarashi, K., et al, 2013. T_{reg} induction by a rationally selected mixture of Clostridia strains from the human microbiota. *Nature*, v. 500, pp. 232–236.

Blaser, M. J. 2006. Who are we? Indigenous microbes and the ecology of human diseases. *EMBO Reports*, v. 7, pp. 956–960.

Blaser, M. J. 2014. *Missing Microbes: How the Overuse of Antibiotics Is Fueling Our Modern Plagues*. New York: Henry Holt & Co., 273 pp.

Cho, I., and Blaser, M. J. 2012. The human microbiome: At the interface of health and disease. *Nature Reviews Genetics*, v. 13, pp. 260–270.

Clark, W. 2008. *In Defense of Self*. New York: Oxford University Press, 265 pp.

Coley, W. B. 1893. The treatment of malignant tumors by repeated inoculations of erysipelas: With a report of ten original cases. *American Journal of the Medical Sciences*, v. 105, pp. 487–511.

Conly, J. M., and Stein, K. 1992. The production of menaquinones (vitamin K2) by intestinal bacteria and their role in maintaining coagulation homeostasis. *Progress in Food & Nutrition Science*, v. 16, pp. 307–343.

Dunn, R. R. 2011. *The Wild Life of Our Bodies: Predators, Parasites, and Partners That Shape Who We Are Today*. New York: HarperCollins, 290 pp.

Ericsson, A. C.; Hagan, C. E.; Davis, D. J.; and Franklin, C. L. 2014. Segmented filamentous bacteria: Commensal microbes with potential effects on research. *Comparative Medicine*, v. 64, pp. 90–98.

Gaboriau-Routhiau, V., et al. 2009. The key role of segmented filamentous bacteria in the coordinated maturation of gut helper T cell responses. *Immunity*, v. 31, pp. 677–689.

Gilbert, S. R.; Sapp, J.; and Tauber, A. I. 2012. A symbiotic view of life: We have never been individuals. *Quarterly Review of Biology*, v. 87, pp. 325–341.

Goodman, A. L., and Gordon, J. I. 2010. Our unindicted coconspirators: Human metabolism from a microbial perspective. *Cell Metabolism*, v. 12, pp. 111–116.

Hold, G. L. 2014. Western lifestyle: A "master" manipulator of the intestinal microbiota? *Gut*, v. 63, pp. 5–6.

Ivanov, I. I., and Honda, K. 2012. Intestinal commensal microbes as immune modulators. *Cell Host & Microbe*, v. 12, pp. 496–508.

Ivanov, I., et al. 2009. Induction of intestinal Th17 cells by segmented filamentous bacteria. *Cell*, v. 139, pp. 485–498.

Jonsson, H. 2013. Segmented filamentous bacteria in human ileostomy samples after high-fiber intake. *FEMS Microbiology Letters*, v. 342, pp. 24–29.

Konkel, L. 2013. The environment within: Exploring the role of the gut microbiome in health and disease. *Environmental Health Perspectives*, v. 121, pp. A276–A281.

Lathrop, S. K., et al. 2011. Peripheral education of the immune system by colonic commensal microbiota. *Nature*, v. 478, pp. 250–254.

LeBlanc, J. G., et al. 2013. Bacteria as vitamin suppliers to their host: A gut microbiota perspective. *Current Opinion in Biotechnology*, v. 24, pp. 160–168.

Lee, S. M., et al. 2013. Bacterial colonization factors control specificity and stability of the gut microbiota. *Nature*, v. 501, pp. 426–429.

Lee, Y. K., and Mazmanian, S. K. 2010. Has the microbiota played a critical role in the evolution of the adaptive immune system? *Science*, v. 330, pp. 1768–1773.

Levine, D. B. 2008. The hospital for the ruptured and crippled: William Bradley Coley, Third Surgeon-in-Chief 1925–1933. *HSS Journal*, v. 4, pp. 1–9.

Lieberman, D. E. 2013. *The Story of the Human Body: Evolution, Health, and Disease.* New York: Pantheon Books, 460 pp.

Maynard, C. L.; Elson, C. O.; Hatton, R. D.; and Weaver, C. T. 2012. Reciprocal interactions of the intestinal microbiota and immune system. *Nature*, v. 489, pp. 231–241.

Mazmanian, S. K., and Kasper, D. L. 2006. The love-hate relationship between bacterial polysaccharides and the host immune system. *Nature Reviews Immunology*, v. 6, pp. 849–858.

Mazmanian, S. K.; Liu, C. H.; Tzianabos, A. O.; and Kasper, D. L. 2005. An immunomodulatory molecule of symbiotic bacteria directs maturation of the host immune system. *Cell*, v. 122, pp. 107–118.

Mazmanian, S. K.; Round, J. L.; and Kasper, D. L. 2008. A microbial symbiosis factor prevents intestinal inflammatory disease. *Nature*, v. 453, pp. 620–625.

McFall-Ngai, M. 2007. Care for the community. *Nature*, v. 445, p. 153.

McFall-Ngai, M. 2008. Are biologists in "future shock"?: Symbiosis integrates biology across domains. *Nature Reviews Microbiology*, v. 6, pp. 789–792.

McFall-Ngai, M., et al. 2013. Animals in a bacterial world: A new imperative for the life sciences. *Proceedings of the National Academy of Sciences*, v. 110, pp. 3229–3236.

Medzhitov, R. 2007. Recognition of microorganisms and activation of the immune response. *Nature*, v. 449, pp. 819–826.

Nicholson, J. K., et al. 2012. Host-gut microbiota metabolic interactions. *Science*, v. 336, pp. 1262–1267.

Rook, G. A. W., and Brunet, L. R. 2002. Give us this day our daily germs. *Biologist*, v. 49, pp. 145–149.

Round, J. L., and Mazmanian, S. K. 2010. Inducible Foxp3^{+} regulatory T-cell development by a commensal bacterium of the intestinal microbiota. *Proceedings of the National Academy of Sciences*, v. 107, pp. 12,204–12,209.

Round, J. L.; O'Connell, R. M.; and Mazmanian, S. K. 2010. Coodination of tolerogenic immune responses by the commensal microbiota. *Journal of Autoimmunity*, v. 34, pp. J220–J225.

Sachs, J. S. 2007. *Good Germs, Bad Germs: Health and Survival in a Bacterial World.* New York: Hill & Wang, 290 pp.

Smith, H. F., et al. 2009. Comparative anatomy and phylogenetic distribution of the mammalian cecal appendix. *Journal of Evolutionary Biology*, v. 22, pp. 1984–1999.

Steinman, R. M., and Cohn, Z. A. 1973. Identification of a novel cell type in peripheral lymphoid organs of mice. I. Morphology, quantification, tissue distribution. *Journal of Experimental Medicine*, v. 137, pp. 1142–1162.

Tauber, A. I. 1994. *The Immune Self: Theory or Metaphor?* Cambridge: Cambridge University Press, 345 pp.

Taylor, L. H.; Latham, S. M.; and Woolhouse, M. E. J. 2001. Risk factors for human disease emergence. *Philosophical Transactions of the Royal Society of London B*, v. 356, pp. 983–989.

Troy, E. B., and Kasper, D. L. 2010. Beneficial effects of *Bacteroides fragilis* polysaccharides on the immune system. *Frontiers in Bioscience*, v. 15, pp. 25–34.

Velasquez-Manoff, M. 2012. *An Epidemic of Absence: A New Way of Understanding Allergies and Autoimmune Diseases.* New York: Scribner, 385 pp.

Wu, H.-J., et al. 2010. Gut-residing segmented filamentous bacteria drive autoimmune arthritis via T helper 17 cells. *Immunity*, v. 32, pp. 815–827.

Yin, Y., et al. 2013. Comparative analysis of the distribution of segmented filamentous bacteria in humans, mice, and chickens. *ISME Journal*, v. 7, pp. 615–621.

CHAPTER 9: INVISIBLE ENEMIES

Blake, J. B. 1952. The Inoculation Controversy in Boston: 1721–1722. *New England Quarterly*, v. 25, pp. 489–506.

Crawford, D. H. 2000. *The Invisible Enemy: A Natural History of Viruses*. Oxford: Oxford University Press, 275 pp.

Crawford, D. H. 2007. *Deadly Companions: How Microbes Shaped Our History*. Oxford: Oxford University Press, 250 pp.

Dixon, B. 1994. *Power Unseen: How Microbes Rule the World*. New York: W. H. Freeman & Co., 237 pp.

Hopkins, D. R. 1983. *Princes and Peasants: Smallpox in History*. Chicago: University of Chicago Press, 380 pp.

Rhodes, J. 2013. *The End of Plagues: The Global Battle Against Infectious Disease*. New York: Palgrave Macmillian, 235 pp.

Riedel, S. 2005. Edward Jenner and the history of smallpox and vaccination. *Baylor University Medical Center Proceedings*, v. 18, pp. 21–25.

Stearns, R. P. 1950. Remarks upon the introduction of inoculation for smallpox in England. *Bulletin of the History of Medicine*, v. 24, pp. 103–122.

Williams, G. 2010. *Angel of Death: The Story of Smallpox*. New York: Palgrave Macmillan, 425 pp.

CHAPTER 10: FEUDING SAVIORS

Centers for Disease Control and Prevention (CDC). 2013. *Antibiotic Resistance Threats in the United States, 2013*. U.S. Department of Health and Human Services, 113 pp.

Crawford, D. H. 2007. *Deadly Companions: How Microbes Shaped Our History*. Oxford: Oxford University Press, 250 pp.

De Kruif, P. 1926. *Microbe Hunters*. New York: Harcourt, Brace, 363 pp.

Dixon, B. 1994. *Power Unseen: How Microbes Rule the World*. New York: W. H. Freeman & Co., 237 pp.

Fleming, A. 1929. On the antibacterial action of cultures of a penicillium, with special reference to their use in isolation of *B. influenza*. *British Journal of Experimental Pathology*, v. 10, pp. 226–236.

Hagar, T. 2006. *The Demon Under the Microscope: From Battlefield Hospitals to Nazi Labs, One Doctor's Heroic Search for the World's First Miracle Drug*. New York: Harmony Books, 340 pp.

Hopwood, D. A. 2007. *Streptomyces in Nature and Medicine: The Antibiotic Makers*. Oxford: Oxford University Press, 250 pp.

Jones, D. S.; Podolsky, S. H.; and Greene, J. A. 2012. The burden of disease and the changing task of medicine. *New England Journal of Medicine*, v. 366, pp. 2333–2338.

Keans, S. 2010. *The Disappearing Spoon: And Other True Tales of Madness, Love, and the History of the World from the Periodic Table of the Elements*. New York: Little Brown, 391 pp.

Lax, E. 2004. *The Mold in Dr. Florey's Coat: The Story of the Penicillin Miracle*. New York: Henry Holt & Co., 307 pp.

Ling, L. L., et al. 2015. A new antibiotic kills pathogens without detectable resistance. *Nature*, v. 517, pp. 455–459.

McKenna, M. 2010. *Superbug: The Fatal Menace of MRSA*. New York: Free Press, 271 pp.

Morgun, A., et al. 2015. Uncovering effects of antibiotics on the host and microbiota using transkingdom gene networks. *Gut*, doi: 10.1136/gutjnl-2014-308820.

Pringle, P. 2012. *Experiment Eleven: Dark Secrets Behind the Discovery of a Wonder Drug*. New York: Walker & Co., 278 pp.

Rhodes, J. 2013. *The End of Plagues: The Global Battle Against Infectious Disease*. New York: Palgrave Macmillian, 235 pp.

Ullmann, A. 2007. Pasteur–Koch: Distinctive ways of thinking about infectious diseases. *Microbe*, v. 2, pp. 383–387.

Vallery-Radot, R. 1926. *The Life of Pasteur*. Garden City, N.Y.: Doubleday, Page & Co., 484 pp.

Williams, G. 2010. *Angel of Death: The Story of Smallpox*. New York: Palgrave Macmillan, 425 pp.

Zimmer, C. 2008. *Microcosm: E. coli and the New Science of Life*. New York: Pantheon, 243 pp.

CHAPTER 11: PERSONAL ALCHEMISTS

Bäckhed, F.; Manchester, J. K.; Semenkovich, C. F.; Gordon, J. I. 2007. Mechanisms underlying the resistance to diet-induced obesity in germ-free mice. *Proceedings of the National Academy of Sciences*, v. 104, pp. 979–984.

Bertola, A., et al. 2012. Identification of adipose tissue dendritic cells correlated with obesity-associated insulin-resistance and inducing Th17 responses in mice and patients. *Diabetes*, v. 61, pp. 2238–2247.

Bulcão, C.; Ferreira, S. R. G.; Giuffrid, F. M. A.; and Ribeiro-Filho, F. F. 2006. The new adipose tissue and adipocytokines. *Current Diabetes Reviews*, v. 2, pp. 19–28.

Cani, P. D. 2012. Crosstalk between the gut microbiota and the endocannabinoid system: Impact on the gut barrier function and the adipose tissue. *Clinical Microbiology and Infection*, v. 4, supplement 4, pp. 50–53.

Cani, P. D., et al. 2007. Metabolic endotoxemia initiates obesity and insulin resistance. *Diabetes*, v. 56, pp. 1761–1772.

Cani, P. D., et al. 2007. Selective increases of bifidobacteria in gut microflora improve high-fat-diet-induced diabetes in mice through a mechanism associated with endotoxaemia. *Diabetologia*, v. 50, pp. 2374–2383.

Cani, P. D., et al. 2008. Changes in gut microbiota control metabolic endotoxemia-induced inflammation in high-fat-diet-induced obesity and diabetes in mice. *Diabetes*, v. 57, pp. 1470–1481.

den Besten, G., et al. 2013. The role of short-chain fatty acids in the interplay between diet, gut microbiota, and host energy metabolism. *Journal of Lipid Research*, v. 54, pp. 2325–2340.

Di Sabatino, A., et al. 2005. Oral butyrate for mildly to moderately active Crohn's disease. *Alimentary Pharmacology & Therapeutics*, v. 22, pp. 789–794.

Duncan, S. H., et al. 2007. Reduced dietary intake of carbohydrates by obese subjects results in decreased concentrations of butyrate and butyrate-producing bacteria in feces. *Applied and Environmental Microbiology*, v. 73, pp. 1073–1078.

Duncan, S. H.; Louis, P.; and Flint, H. J. 2004. Lactate-utilizing bacteria, isolated from human feces, that produce butyrate as a major fermentation product. *Applied Environmental Microbiology*, v. 70, pp. 5810–5817.

El Kaoutari, et al. 2013. The abundance and variety of carbohydrate-active enzymes in the human gut microbiota. *Nature Reviews Microbiology*, v. 11, pp. 497–504.

Fei, N., and Zhao, L. 2013. An opportunistic pathogen isolated from the gut of an obese human causes obesity in germfree mice. *ISME Journal*, v. 7, pp. 880–884.

Fukuda, S., et al. 2011. Bifidobacteria can protect from enteropathogenic infection through production of acetate. *Nature*, v. 469, pp. 543–547.

Furusawa, Y., et al. 2013. Commensal microbe-derived butyrate induces the differentiation of colonic regulatory T cells. *Nature*, v. 504, pp. 446–450.

Gourko, H.; Williamson, D. I.; and Tauber, A. I. 2000. *The Evolutionary Biology Papers of Elie Metchnikoff.* Dordrecht, Boston, and London: Kluwer Academic Publishers, 221 pp.

Harig, J. M.; Soergel, K. H.; Komorowski, R. A.; and Wood, C. M. 1989. Treatment of diversion colitis with short-chain-fatty-acid irrigation. *New England Journal of Medicine*, v. 320, pp. 23–28.

Hullar, M. A. J.; Burnett-Hartman, A. N.; and Lampe, J. W. 2014. Gut microbes, diet, and cancer. In *Advances in Nutrition and Cancer.* Ed. V. Zappia et al. *Cancer Treatment and Research*, v. 159. Berlin and Heidelberg: Springer Verlag, pp. 377–399.

Hvistendahl, M. 2012. My microbiome and me. *Science*, v. 336, pp. 1248–1250.

Kau, A. L., et al. 2011. Human nutrition, the gut microbiome, and the immune system. *Nature*, v. 474, pp. 327–336.

Kuo, S.-M. 2013. The interplay between fiber and the intestinal microbiome in the inflammatory response. *Advances in Nutrition*, v. 4, pp. 16–28.

Ley, R. E., et al. 2006. Microbial ecology: Human gut microbes associated with obesity. *Nature*, v. 444, pp. 1022–1023.

Mackenbach, J. P., and Looman, C. W. N. 2013. Life expectancy and national income in Europe, 1900–2008: An update of Preston's analysis. *International Journal of Epidemiology*, v. 42, pp. 1100–1110.

McLaughlin, T., et al. 2014. T-cell profile in adipose tissue is associated with insulin resistance and systemic inflammation in humans. *Arteriosclerosis, Thrombosis, and Vascular Biology*, v. 34, pp. 2637–2643.

Metchnikoff, É. 1908. *The Prolongation of Life: Optimistic Studies.* Trans. P. C. Mitchell. New York and London: Knickerbocker Press, G. P. Putnam's Sons, 343 pp.

Metchnikoff, O. 1921. *Life of Elie Metchnikoff, 1845–1916.* Boston and New York: Houghton Mifflin, 297 pp.

Podolsky, S. 1998. Cultural divergence: Elie Metchnikoff's *Bacillus bulgaricus* therapy and his underlying concept of health. *Bulletin of the History of Medicine*, v. 72, pp. 1–27.

Podolsky, S. H. 2012. The art of medicine: Metchnikoff and the microbiome. *The Lancet*, v. 380, pp. 1810–1811.

Ridaura, V. K., et al. 2013. Gut microbiota from twins discordant for obesity modulate metabolism in mice. *Science*, v. 341, 1241214, doi:10.1126/science.1241214.

Ridlon, J. M.; Kang, D. J.; and Hylemon, P. B. 2006. Bile salt biotransformations by human intestinal bacteria. *Journal of Lipid Research*, v. 47, pp. 241–259.

Roy, C. C.; Kien, C. L.; Bouthillier, L.; and Levy, E. 2006. Short-chain fatty acids: Ready for prime time? *Nutrition in Clinical Practice*, v. 21, pp. 351–366.

Scheppach, W., et al. 1992. Effect of butyrate enemas on the colonic mucosa in distal ulcerative colitis. *Gastroenterology*, v. 103, pp. 51–56.

Sekirov, I.; Russell S. L.; Antunes, L. C. M.; and Finlay, B. B. 2010. Gut microbiota in health and disease. *Physiological Reviews*, v. 90, pp. 859–904.

Singh, N., et al. 2014. Activation of Gpr109a, receptor for niacin and the commensal metabolite butyrate, suppresses colonic inflammation and carcinogenesis. *Immunity*, v. 40, pp. 128–139.

Smith P. M., et al. 2013. The microbial metabolites, short-chain fatty acids, regulate colonic Treg cell homeostasis. *Science*, v. 341 pp. 569–573.

Surmi, B. K., and Hasty, A. H. 2008. Macrophage infiltration into adipose tissue: Initiation, propagation, and remodeling. *Future Lipidology*, v. 3, pp. 545–556.

Taubes, T. 2009. Prosperity's plague. *Science*, v. 325, pp. 256–260.

Tilg, H., and Kaser, A. 2011. Gut microbiome, obesity, and metabolic dysfunction. *Journal of Clinical Investigation*, v. 121, pp. 2126–2132.

Tremaroli, V., and Bäckhed, F. 2012. Functional interactions between the gut microbiota and host metabolism. *Nature*, v. 489, pp. 242–249.

van Immerseel, F., et al. 2010. Butyric acid-producing anaerobic bacteria as a novel probiotic treatment approach for inflammatory bowel disease. *Journal of Medical Microbiology*, v. 59, pp. 141–143.

Vrieze, A., et al. 2012. Transfer of intestinal microbiota from lean donors increases insulin sensitivity in individuals with metabolic syndrome. *Gastroenterology*, v. 143, pp. 913–916.

Walker, A. W., and Parkhill, J. 2013. Fighting obesity with bacteria. *Science*, v. 341, p. 1069.

Wellen, K. E., and Hotamisligil, G. S. 2005. Inflammation, stress, and diabetes. *Journal of Clinical Investigation*, v. 115, pp. 1111–1119.

Wong, J. M., et al. 2006. Colonic health: Fermentation and short-chain fatty acids. *Journal of Clinical Gastroenterology*, v. 40, pp. 235–243.

Xiao, S., et al. 2014. A gut microbiota-targeted dietary intervention for amelioration of chronic inflammation underlying metabolic syndrome. *FEMS Microbiology Ecology*, v. 87, pp. 357–367.

Zhang, C., et al. 2010. Interactions between gut microbiota, host genetics, and diet relevant to development of metabolic syndromes in mice. *ISME Journal*, v. 4, pp. 232–241.

Zhang, C., et al. 2012. Structural resilience of the gut microbiota in adult mice under high-fat dietary perturbations. *ISME Journal*, v. 6, pp. 1848–1857.

Zhao, L. 2013. The gut microbiota and obesity: From correlation to causality. *Nature*, v. 11, pp. 639–647.

Zoetendal, E. G., et al. 2012. The human small intestinal microbiota is driven by rapid uptake and conversion of simple carbohydrates. *ISME Journal*, v. 6, pp. 1415–1426.

CHAPTER 12: TENDING THE GARDEN

Anukam, K. C., et al. 2006. Augmentation of antimicrobial metronidazole therapy of bacterial vaginosis with oral *Lactobacillus rhamnosus* GR-1 and *Lactobacillus reuteri* RC-14: Randomized, double-blind, placebo-controlled trial. *Microbes and Infection*, v. 8, pp. 1450–1454.

Anukam, K. C., et al. 2006. Clinical study comparing *Lactobacillus* GR-1 and RC-14 with metronidazole vaginal gel to treat symptomatic bacterial vaginosis. *Microbes and Infection*, v. 8, pp. 2772–2776.

Atassi, F., and Servin, A. L. 2010. Individual and co-operative roles of lactic acid and hydrogen peroxide in the killing activity of enteric strain Lactobacillus johnsonii NCC933 and vaginal strain Lactobacillus gasseri KS120.1 against enteric, uropathogenic, and vaginosis-associated pathogens. *FEMS Microbiology Letters*, v. 304, pp. 29–38.

Bajaj, J. S., et al. 2014. Randomized clinical trial: Lactobacillus GG modulates gut microbiome, metabolome, and endotoxemia in patients with cirrhosis. *Alimentary Pharmacology and Therapeutics*, v. 39, pp. 1113–1125.

Barrett, J. S. 2013. Extending our knowledge of fermentable, short-chain carbohydrates for managing gastrointestinal symptoms. *Nutrition in Clinical Practice*, v. 28, pp. 300–306.

Barrons, R., and Tassone, D. 2008. Use of *Lactobacillus* probiotics for bacterial genitourinary infections in women: A review. *Clinical Therapeutics*, v. 30, pp. 453–468.

Bermudez-Brito, M., et al. 2012. Probiotic mechanisms of action. *Annals of Nutrition and Metabolism*, v. 61, pp. 160–174.

Bernstein, A. M., et al. 2013. Major cereal grain fibers and psyllium in relation to cardiovascular health. *Nutrients*, v. 5, pp. 1471–1487.

Brandt, L. J., and Aroniadis, O. C. 2013. An overview of fecal microbiota transplantation: Techniques, indications, and outcomes. *Gastrointestinal Endoscopy*, v. 78, pp. 240–249.

Clemens, R., et al. 2012. Filling America's fiber intake gap: Summary of a roundtable to probe realistic solutions with a focus on grain-based foods. *Journal of Nutrition*, v. 142, pp. 1390S–1401S.

David, L. A., et al. 2014. Diet rapidly and reproducibly alters the human gut microbiome. *Nature*, v. 505, pp. 559–563.

Delzenne, N.; Neyrinck, A. M.; Bäckhed, F.; and Cani, P. D. 2011. Targeting gut microbiota in obesity: Effects of prebiotics and probiotics. *Nature Reviews Endocrinology*, v. 7, pp. 639–646.

Delzenne, N.; Neyrinck, A. M.; and Cani, P. D. 2013. Gut microbiota and metabolic disorders: How prebiotic can work? *British Journal of Nutrition*, v. 109, pp. S81–S85.

Eiseman, B., et al. 1958. Fecal enema as an adjunct in the treatment of pseudomembranous enterocolitis. *Surgery*, v. 44, pp. 854–859.

El Kaoutari, A., et al. 2013. The abundance and variety of carbohydrate-active enzymes in the human gut microbiota. *Nature Reviews Microbiology*, v. 11, pp. 497–504.

Falagas, M. E.; Betsi, G. I.; and Athanasiou, S. 2007. Probiotics for the treatment of women with bacterial vaginosis. *Clinical Microbiology and Infection*, v. 13, pp. 657–664.

Gough, E.; Shaikh, H.; and Manges, A. R. 2011. Systematic review of intestinal microbiota transplantation (fecal bacteriotherapy) for recurrent *Clostridium difficile* infection. *Clinical Infectious Diseases*, v. 53, pp. 994–1002.

Gross, L. S.; Li, L.; Ford, E. S.; and Lui, S. 2004. Increased consumption of refined carbohydrates and the epidemic of type 2 diabetes in the United States: An ecologic assessment. *American Journal of Clinical Nutrition*, v. 79, pp. 774–779.

Hill, C., and Sanders, M. E. 2013. Rethinking "probiotics." *Gut Microbes* v. 4, pp. 269–270.

Hume, M. E. 2011. Historic perspective: Prebiotics, probiotics, and other alternatives to antibiotics. *Poultry Science*, v. 90, pp. 2663–2669.

Kassam, Z.; Lee, C. H.; Yuan, Y.; and Hunt, R. H. 2013. Fecal microbiota transplantation for *Clostridium difficile* infection: Systematic review and meta-analysis. *American Journal of Gastorenterology*, v. 108, pp. 500–508.

Kelly, C. P. 2013. Fecal microbiota transplantation: An old therapy comes of age. *New England Journal of Medicine*, v. 368, pp. 474–475.

Khoruts, A.; Dicksved, J.; Jansson, J. K.; and Sadowsky, M. H. 2010. Changes in the composition of the human fecal microbiome after bacteriotherapy for recurrent *Clostridium difficile*-associated diarrhea. *Journal of Clinical Gastroenterology*, v. 44, pp. 354–360.

Korecka, A., and Arulampalam, V. 2012. The gut microbiome: Scourge, sentinel, or spectator? *Journal of Oral Microbiology*, v. 4, p. 9367, doi:10.3402/jom.v4i0.9367.

Kumar, V., et al. 2012. Dietary roles of non-starch polysaccharides in human nutrition: A review. *Critical Reviews in Food Science and Nutrition*, v. 52, pp. 899–935.

Lemon, K. P.; Armitage, G. C.; Relman, D. A.; and Fischbach, M. A. 2012. Microbiota-targeted therapies: An ecological perspective. *Science Translational Medicine*, v. 4, p. 137rv5.

Ling, Z., et al. 2013. The restoration of the vaginal microbiota after treatment for bacterial vaginosis with metronidazole or probiotics. *Microbial Ecology*, v. 65, pp. 773–780.

Macfarlane, G. T., and Macfarlane, S. 2011. Fermentation in the human large intestine: Its physiologic consequences and the potential contribution of prebiotics. *Journal of Clinical Gastroenterology*, v. 45, pp. S120–S127.

MacPhee, R. A., et al. 2010. Probiotic strategies for the treatment and prevention of bacterial vaginosis. *Expert Opinion on Pharmacotherapy*, v. 11, pp. 2985–2995.

Mastromarino, P.; Vitali, B.; and Mosca, L. 2013. Bacterial vaginosis: A review on clinical trials with probiotics. *New Microbiologica*, v. 36, pp. 229–238.

Mirmonsef, P., et al. 2014. Free glycogen in vaginal fluids is associated with Lactobacillus colonization and low vaginal pH. *PLOS ONE*, v. 9, p. e102467.

O'Keefe, S. J., et al. 2009. Products of the colonic microbiota mediate the effects of diet on colon cancer risk. *Journal of Nutrition*, v. 139, pp. 2044–2048.

Petrof, E. O., et al. 2013. Stool substitute transplant therapy for the eradiation of *Clostridium difficile* infection: "RePOOPulating" the gut. *Microbiome*, v. 1, p. 3.

Rastall, R. A., and Gibson, G. R. 2015. Recent developments in prebiotics to selectively impact beneficial microbes and promote intestinal health. *Current Opinion in Biotechnology*, v. 32, pp. 42–46.

Reid, G.; Jass, J.; Sebulsky, M. T.; and McCormick, J. K. 2003. Potential uses of probiotics in clinical practice. *Clinical Microbiology Reviews*, v. 16, pp. 658–672.

Reid, G., et al. 2003. Oral use of *Lactobacillus rhamnosus* GR-1 and *L. fermentum* RC-14 significantly alters vaginal flora: Randomized, placebo-controlled trial in 64 healthy women. *FEMS Immunology and Medical Microbiology*, v. 35, pp. 131–134.

Ritchie, M. L., and Romanuk, T. N. 2012. A meta-analysis of probiotic efficacy for gastrointestinal diseases. *PLOS ONE*, v. 7, p. e34938. doi:10.1371/journal.pone.0034938.

Roberfroid, M. 2007. Prebiotics: The concept revisited. *Journal of Nutrition*, v. 137, pp. 830S–837S.

Roberfroid, M., et al. 2010. Prebiotic effects: Metabolic and health benefits. *British Journal of Nutrition*, v. 104, supp. 2, pp. S1-S63.

Rohlke, F.; Surawicz, C. M.; and Stollman, N. 2010. Fecal flora reconstitution for recurrent *Clostridium difficile* infection: Results and methodology. *Journal of Clinical Gastroenterology*, v. 44, pp. 567–570.

Russell, W. R., et al. 2011. High-protein, reduced-carbohydrate weight-loss diets promote metabolite profiles likely to be detrimental to colonic health. *American Journal of Clinical Nutrition*, v. 5, pp. 1062–1072.

Sears, C. L., and Garrett, W. S. 2014. Microbes, microbiota, and colon cancer. *Cell Host & Microbe*, v. 17, pp. 317–328.

Shankar, V., et al. 2014. Species and genus level resolution analysis of gut microbiota in *Clostridium difficile* patients following fecal microbiota transplantation. *Microbiome*, v. 2, p. 13.

Smith, M. B.; Kelly, C.; and Alm, E. J. 2014. How to regulate faecal transplants. *Nature*, v. 506, pp. 290–291.

Song, Y., et al. 2013. Microbiota dynamics in patients treated with fecal microbiota transplantation for recurrent *Clostridium difficile* infection. *PLOS ONE*, v. 8, p. e81330, doi:10.1371/journal.pone.0081330.

Surawicz, C. M., and Alexander, J. 2011. Treatment of refractory and recurrent *Clostridium difficile* infection. *Nature Reviews Gastroenterology*, v. 8, pp. 330–339.

Talbot, H. K., et al. 2011. Effectiveness of season vaccine in preventing confirmed influenza-associated hospitalizations in community dwelling older adults. *Journal of Infectious Disease*, v. 203, pp. 500–508.

van Nood, E., et al. 2013. Duodenal infusion of donor feces for recurrent *Clostridium difficile*. *New England Journal of Medicine*, v. 368, pp. 407–415.

Vipperla, K., and O'Keefe, S. J. 2012. The microbiota and its metabolites in colonic mucosal health and cancer risk. *Nutrition in Clinical Practice*, v. 27, pp. 624–635.

Walter, J., and Ley, R. 2011. The human gut microbiome: Ecology and recent evolutionary changes. *Annual Review of Microbiology*, v. 65, pp. 411–429.

Wang, J., et al. 2015. Modulation of gut microbiota during probiotic-mediated attenuation of metabolic syndrome in high-fat-diet-fed mice. *ISME Journal*, v. 9, pp. 1–15.

Wilson, M. 2008. *Bacteriology of Humans: An Ecological Perspective*. Malden, Mass.; Oxford, U.K.; Victoria, Austral.: Blackwell Publishing, 351 pp.

CHAPTER 13: COURTING ANCIENT FRIENDS

Ackermann, W., et al. 2015. The influence of glyphosate on the microbiota and production of botulinum neurotoxin during ruminal fermentation. *Current Microbiology*, v. 70, pp. 374–382.

Adesemoye, A. O.; Torbert, H. A.; and Kloepper, J. W. 2009. Plant growth-promoting rhizobacteria allow reduced application rates of chemical fertilizers. *Microbial Ecology*, v. 58, pp. 921–929.

Ahemad, M., and Khan, M. S. 2011. Toxicological effects of selective herbicides on plant growth promoting activities of phosphate solubilizing *Klebsiella* sp. strain PS19. *Current Microbiology*, v. 62, pp. 532–538.

Albrecht, W. A. 1938. Loss of soil organic matter and its restoration In *Soils and Men*, Yearbook of Agriculture, U.S. Department of Agriculture. Washington, D.C.: U.S. Government Printing Office, pp. 347–360.

Albrecht, W. A. 1939. Variable levels of biological activity in Sanborn Field after fifty years of treatment. *Soil Science Society of America Proceedings*, v. 3, pp. 77–82.

Albrecht, W. A. 1947. Our teeth and our soil. *Annals of Dentistry*, v. 8, no. 4 (December), pp. 199–213.

Alloway, B. J., ed. 2008. *Micronutrient Deficiencies in Global Crop Production*. Heidelberg: Springer, 353 pp.

Baig, K., et al. 2012. Comparative effectiveness of *Bacillus* spp. possessing either dual or single growth-promoting traits for improving phosphorus uptake, growth, and yield of wheat (*Triticum aestivum* L.). *Annals of Microbiology*, v. 62, pp. 1109–1119.

Bais, H. P., et al. 2005. Mediation of pathogen resistance by exudation of antimicrobials from roots. *Nature*, v. 434, pp. 217–221.

Balemi, T., and Negisho, K. 2012. Management of soil phosphorus and plant adaptation mechanisms to phosphorus stress for sustainable crop production: A review. *Journal of Soil Science and Plant Nutrition*, v. 12, pp. 547–561.

Balfour, E. B. 1943. *The Living Soil: Evidence of the Importance to Human Health of Soil Vitality, with Special Reference to National Planning*. London: Faber & Faber, 246 pp.

Beauregard, P. B., et al. 2013. *Bacillus subtilis* biofilm induction by plant polysaccharides. *Proceedings of the National Academy of Sciences*, v. 110, pp. E1621–E1630.

Belimov, A. A.; Kojemiakov, A. P.; and Chuvarliyeva, C. V. 1995. Interaction between barley and mixed cultures of nitrogen fixing and phosphate-solubilizing bacteria. *Plant and Soil*, v. 173, pp. 29–37.

Berendsen, R. L.; Pieterse, C. M. J.; and Bakker, P. A. H. M. 2012. The rhizosphere microbiome and plant health. *Trends in Plant Science*, v. 17, pp. 478–486.

Birkhofer, K., et al. 2008. Long-term organic farming fosters below and aboveground biota: Implications for soil quality, biological control, and productivity. *Soil Biology & Biochemistry*, v. 40, pp. 2297–2308.

Bloemberg, G. V., and Lugtenberg, B. J. J. 2001. Molecular basis of plant growth promotion and biocontrol by rhizobacteria. *Current Opinion in Plant Biology*, v. 4, pp. 343–350.

Bøhn, T., et al. 2014. Compositional differences in soybeans on the market: Glyphosate accumulates in Roundup Ready GM soybeans. *Food Chemistry*, v. 153, pp. 207–215.

Cordell, D.; Rosemarin, A.; Schröder, J. J.; and Smit, A. L. 2011. Towards global phosphorus security: A systems framework for phosphorus recovery and reuse options. *Chemosphere*, v. 84, pp. 747–758.

Cushnie, T. P. T., and Lamb, A. J. 2005. Antimicrobial activity of flavonoids. *International Journal of Antimicrobial Agents*, v. 26, pp. 343–356.

Daldy, Y. 1940. Food production without artificial fertilizers. *Nature*, v. 145, pp. 905–906.

Davis, D. R. 2009. Declining fruit and vegetable nutrient composition: What is the evidence? *HortScience*, v. 44, pp. 15–19.

Davis, D.; Epp, M.; and Riordan, H. 2004. Changes in USDA food composition data for 43 garden crops, 1950–1999. *Journal of the American College of Nutrition*, v. 23, pp. 669–682.

Dennis, P. G.; Miller, A. J.; and Hirsch, P. R. 2010. Are root exudates more important than other sources of rhizodeposits in structuring rhizosphere bacterial communities? *FEMS Microbiology Ecology*, v. 72, pp. 313–327.

Farrar, J.; Hawes, M.; Jones, D.; and Lindow, S. 2003. How roots control the flux of carbon to the rhizosphere. *Ecology*, v. 84, pp. 827–837.

Gaiero, J. R., et al. 2013. Inside the root microbiome: Bacterial root endophytes and plant growth promotion. *American Journal of Botany*, v. 100, pp. 1738–1750.

Garbaye, J. 1994. Helper bacteria: A new dimension to the mycorrhizal symbiosis. *New Phytologist*, v. 128, pp. 197–210.

Garvin, D. F.; Welch, R. M.; and Finley, J. W. 2006. Historical shifts in the seed mineral micronutrient concentration of US hard red winter wheat germplasm. *Journal of the Science of Food and Agriculture*, v. 86, pp. 2213–2220.

Glick, B. R. 1995. The enhancement of plant growth by free-living bacteria. *Canadian Journal of Microbiology*, v. 41, pp. 109–117.

Goldstein, A. H.; Rogers, R. D.; and Mead, G. 1993. Mining by microbe. *BioTechnology*, v. 11, pp. 1250–1254.

Hameeda, B., et al. 2008. Growth promotion of maize by phosphate solubilizing bacteria isolated from composts and macrofauna. *Microbiologial Research*, v. 163, pp. 234–242.

Herr, I., and Büchler, M. W. 2010. Dietary constituents of broccoli and other cruciferous vegetables: Implications for prevention and therapy of cancer. *Cancer Treatment Reviews*, v. 36, pp. 377–383.

Hoitink, H., and Boehm, M. 1999. Biocontrol within the context of soil microbial communities: A substrate-dependent phenomenon. *Annual Review of Phytopathology*, v. 37, pp. 427–446.

Hong, H. A., et al. 2009. *Bacillus subtilis* isolated from the human gastrointestinal tract. *Research in Microbiology*, v. 160, pp. 134–143.

Howard, A. 1939. Medical "testament" on nutrition. *British Medical Journal*, v. 1, p. 1106.

Howard, A. 1940 (1945). *An Agricultural Testament*. London, New York, and Toronto: Oxford University Press, 253 pp.

Jarrell, W. M., and Beverly, R. B. 1981. The dilution effect in plant nutrient studies. *Advances in Agronomy*, v. 34, pp. 197–224.

Jones, D. L.; Nguyen, C.; and Finlay, R. D. 2009. Carbon flow in the rhizosphere: Carbon trading at the soil-root interface. *Plant and Soil*, v. 321, pp. 5–33.

Jones, D. L., et al. 2013. Nutrient stripping: The global disparity between food security and soil nutrient stocks. *Journal of Applied Ecology*, v. 50, pp. 851–862.

Kaempffert, W. 1940. Science in the news. *New York Times*, June 30, 1940, p. 41.

Khan, M. S.; Zaidi, A.; and Wani, P. A. 2007. Role of phosphate-solubilizing microorganisms in sustainable agriculture: A review. *Agronomy and Sustainable Development*, v. 27, pp. 29–43.

Khan, S. A., et al. 2007. The myth of nitrogen fertilization for soil carbon sequestration. *Journal of Environmental Quality*, v. 36, pp. 1821–1832.

Kloepper, J. W.; Lifshitz, K.; Zoblotowicz, R. M. 1989. Free-living bacterial inocula for enhancing crop productivity. *Trends in Biotechnology*, v. 7, pp. 39–43.

Knekt, P., et al. 2002. Flavonoid intake and risk of chronic diseases. *American Journal of Clinical Nutrition*, v. 76, pp. 560–568.

Krüger, M.; Shehata, A. A.; Schrödl, W.; and Rodloff, A. 2013. Glyphosate suppresses the antagonistic effect of *Enterococcus* spp. on *Clostridium botulinum*. *Anaerobe*, v. 20, pp. 74–78.

Kucey, R. M. N.; Janzen, H. H.; and Leggett, M. E. 1989. Microbially mediated increases in plant-available phosphorus. *Advances in Agronomy*, v. 42, pp. 199–228.

Lasat, M. M. 2002. Phytoextraction of toxic metals: A review of biological mechanisms. *Journal of Environmental Quality*, v. 31, pp. 109–120.

Lee, B.; Lee, S.; and Ryu, C. M. 2012. Foliar aphid feeding recruits rhizosphere bacteria and primes plant immunity against pathogenic and nonpathogenic bacteria in pepper. *Annals of Botany*, v. 110, pp. 281–290.

López-Guerrero, M. G., et al. 2013. Buffet hypothesis for microbial nutrition at the rhizosphere. *Frontiers in Plant Science*, v. 4, pp. 1–4.

Marschner, H., and Dell, B. 1994. Nutrient uptake in mycorrhizal symbiosis. *Plant and Soil*, v. 159, pp. 89–102.

Mayer, A. M. 1997. Historical changes in the mineral content of fruits and vegetables. *British Food Journal*, v. 99, pp. 207–211.

Mendes, R., et al. 2011. Deciphering the rhizosphere microbiome for disease-suppressive bacteria. *Science*, v. 332, pp. 1097–1100.

Miller, D. D., and Welch, R. M. 2013. Food system strategies for preventing micronutrient malnutrition. *Food Policy*, v. 42, pp. 115–128.

Mulvaney, R. L.; Khan, S. A.; and Ellsworth, T. R. 2009. Synthetic nitrogen fertilizers deplete soil nitrogen: A global dilemma for sustainable cereal production. *Journal of Environmental Quality*, v. 38, pp. 2295–2314.

Neumann, G., et al. 2006. Relevance of glyphosate transfer to non-target plants via the rhizosphere. *Journal of Plant Diseases and Protection*, v. 20, pp. 963–969.

Pachikian, B. D., et al. 2010. Changes in intestinal bifidobacteria levels are associated with the inflammatory response in magnesium-deficient mice. *Journal of Nutrition*, v. 140, pp. 590–514.

Peters, R. D.; Sturz, A. V.; Carter, M. R.; and Sanderson, J. B. 2003. Developing disease-suppressive soils through crop rotation and tillage management practices. *Soil & Tillage Research*, v. 72, pp. 181–192.

Raaijmakers, J. M., et al. 2009. The rhizosphere: A playground and battlefield for soilborne pathogens and beneficial microorganisms. *Plant and Soil*, v. 321, pp. 341–361.

Raghu, K., and MacRae, I. C. 1966. Occurrence of phosphate-dissolving microorganisms in the rhizosphere of rice plants and in submerged soils. *Journal of Applied Bacteriology*, v. 29, pp. 582–586.

Ramesh, R.; Joshi, A.; and Ghanekar, M. P. 2008. *Pseudomonads*: Major antagonistic endophytic bacteria to suppress bacterial wilt pathogen. *Ralstonia solanacearum* in the eggplant (*Solanum memongena* L.). *World Journal of Microbiology & Biotechnology*, v. 25, pp. 47–55.

Ramírez-Puebla, S. T., et al. 2013. Gut and root microbiota commonalities. *Applied and Environmental Microbiology*, v. 79, pp. 2–9.

Ray, J.; Bagyaraj, D. J.; and Manjunath, A. 1981. Influence of soil inoculation with vesicular-arbuscular mycorrihza and a phosphate-dissolving bacterium on plant growth and ^{32}P uptake. *Soil Biology and Biochemistry*, v. 13, pp. 105–108.

Rodríguez, H., and Fraga, R. 1999. Phosphate solubilizing bacteria and their role in plant growth promotion. *Biotechnology Advances*, v. 17, pp. 319–339.

Ryan, M. H.; Derrick, J. W.; and Dann, P. R. 2004. Grain mineral concentrations and yield of wheat grown under organic and conventional management. *Journal of the Science of Food and Agriculture*, v. 84, pp. 207–216.

Ryan, P. R.; Delhaize, E.; and Jones, D. L. 2001. Function and mechanism of organic anion exudation from plant roots. *Annual Review of Plant Physiology and Plant Molecular Biology*, v. 52, pp. 527–560.

Santi, C.; Bogusz, D.; and Franche C. 2013. Biological nitrogen fixation in non-legume plants. *Annals of Botany*, v. 111, pp. 743–767.

Santos, V. B., et al. 2012. Soil microbial biomass and organic matter fractions during transition from conventional to organic farming systems. *Geoderma*, v. 170, pp. 227–231.

Schrödl, W., et al. 2014. Possible effects of glyphosate on *Mucorales* abundance in the rumen of dairy cows in Germany. *Current Microbiology*, v. 69, pp. 817–823.

Seghers, D., et al. 2004. Impact of agricultural practices on the *Zea mays* L. endophytic community. *Applied and Environmental Microbiology*, v. 70, pp. 1475–1482.

Sharma, K. N., and Deb, D. L. 1988. Effect of organic manuring on zinc diffusion in soil of varying texture. *Journal of the Indian Society of Soil Science*, v. 36, pp. 219–224.

Shehata, A. A., et al. 2013. The effect of glyphosate on potential pathogens and beneficial members of poultry microbiota in vitro. *Current Microbiology*, v. 66, pp. 350–358.

Tarafdar, J. C., and Claassen, N. 1988. Organic phosphorus compounds as a phosphorus source for higher plants through the activity of phosphatases produced by plant roots and microorganisms. *Biology and Fertility of Soils*, v. 5, pp. 308–312.

Thomas, D.. 2003. A study on the mineral depletion of the foods available to us as a nation over the period 1940 to 1991. *Nutrition and Health*, v. 17, pp. 85–115.

Thomas, D. 2007. The mineral depletion of foods available to us as a nation (1940–2002): A review of the 6th edition of McCance and Widdowson. *Nutrition and Health*, v. 19, pp. 21–55.

Toro, M.; Azcón, R.; and Barea, J. M. 1997. Improvement of arbuscular mycorrhiza development by inoculation of soil with phosphate-solubilizing rhizobacteria to improve rock phosphate bioavailability (^{32}P) and nutrient cycling. *Applied and Environmental Microbiology*, v. 63, pp. 4408–4412.

Tu, C.; Ristaino, J. B.; and Hu. S. 2006. Soil microbial biomass and activity in organic tomato farming systems: Effects of organic inputs and straw mulching. *Soil Biology & Biochemistry*, v. 38, pp. 247–255.

Tu, C., et al. 2006. Responses of soil microbial biomass and N availability to transition strategies from conventional to organic farming systems. *Agriculture, Ecosystems and Environment*, v. 113, pp. 206–215.

U.S. Department of Agriculture. 2011. *Composition of Foods: Raw, Processes, Prepared.* USDA National Nutrient Database for Standard Reference, Release 24.

Vessey, J. K. 2003. Plant growth promoting rhizobacteria as biofertilizers. *Plant and Soil*, v. 255, pp. 571–586.

Waksman, S. A., and Tenney, F. G. 1928. Composition of natural organic materials and their decomposition in the soil: III. The influence of nature of plant upon the rapidity of its decomposition. *Soil Science*, v. 26, pp. 155–171.

Welbaum, G. E.; Sturz, A. V.; Dong, Z. M.; and Nowak, J. 2007. Managing soil microorganisms to improve productivity of agroecosystems. *Critical Reviews of Plant Science*, v. 23, pp. 175–193.

White, P. J., and Broadley, M. R. 2005. Historical variation in the mineral composition of edible horticultural products. *Journal of Horticultural Science and Biotechnology*, v. 80, pp. 660–667.

White, P. J., and Brown, P. H. 2010. Plant nutrition for sustainable development and global health. *Annals of Botany*, v. 105, pp. 1073–1080.

Yang, J. W., et al. 2011. Whitefly infestation of pepper plants elicits defense responses against bacterial pathogens in leaves and roots and changes the below-ground microflora. *Journal of Ecology*, v. 99, pp. 46–56.

Yazdani, M., and Bahmanyar, M. 2009. Effect of phosphate solubilization microorganisms (PSM) and plant growth promoting rhizobacteria (PGPR) on yield and yield components of corn (*Zea mays* L.). *World Academy of Science, Engineering and Technology*, v. 49, pp. 90–92.

CHAPTER 14: CULTIVATING HEALTH

Balfour Sartor, R. 2008. Microbial influences in inflammatory bowel diseases. *Gastroenterology*, v. 134, pp. 577–594.

Blaser, M. J. 2014. *Missing Microbes: How the Overuse of Antibiotics Is Fueling Our Modern Plagues*. New York: Henry Holt, 273 pp.

Bravo, J. A., et al. 2012. Communication between gastrointestinal bacteria and the nervous system. *Current Opinion in Pharmacology*, v. 12, pp. 667–672.

Collins, S. M.; Surette, M.; and Bercik, P. 2012. The interplay between the intestinal microbiota and the brain. *Nature Reviews Microbiology*, v. 10, pp. 735–742.

Ege, M. J., et al. 2011. Exposure to environmental microorganisms and childhood asthma. *New England Journal of Medicine*, v. 364, pp. 701–709.

Hanski, I., et al. 2012. Environmental biodiversity, human microbiota, and allergy are interrelated. *Proceedings of the National Academy of Sciences*, v. 109, pp. 8334–8339.

Hsiao, E., et al. 2012. Modeling an autism risk factor in mice leads to permanent immune dysregulation. *Proceedings of the National Academy of Sciences*, v. 109, pp. 12,776–12,781.

Hsiao, E., et al. 2013. Microbiota modulate behavioral and physiological abnormalities associated with neurodevelopmental disorders. *Cell*, v. 155, pp. 1451–1463.

Koeth, R. A., et al. 2013. Intestinal microbiota metabolism of L-carnitine, a nutrient in red meat, promotes atherosclerosis. *Nature Medicine*, v. 19, pp. 576–585.

Lee, Y. K., et al. 2011. Proinflammatory T-cell responses to gut microbiota promote experimental autoimmune encephalomyelitis. *Proceedings of the National Academy of Sciences*, v. 108, pp. 4615–4622.

Mayer, E. A., et al. 2014. Gut microbes and the brain: Paradigm shift in neuroscience. *Journal of Neuroscience*, v. 34, pp. 15,490–15,496.

Missaghi, B. 2014. Perturbation of the human microbiome as a contributor to inflammatory bowel disease. *Pathogens*, v. 3, pp. 510–527.

Ochoa-Repáraz, J., et al. 2010. A polysaccharide from the human commensal *Bacteroides fragilis* protects against CNS demyelinating disease. *Mucosal Immunology*, v. 3, pp. 487–495.

Sessitsch, A., and Mitter, B. 2015. 21st century agriculture: Integration of plant microbiomes for improved crop production and food security. *Microbial Biotechnology*, v. 8, pp. 32–33.

Shreiner, A. B.; Kao, J. Y.; and Young, V. B. 2015. The gut microbiome in health and in disease. *Current Opinion in Gastroenterology*, v. 31, pp. 69–75.

Stefka, A. T., et al. 2014. Commensal bacteria protect against food allergen sensitization. *Proceedings of the National Academy of Sciences*, v. 111, pp. 12,145–13,150.

Suez, J., et al. 2014. Artificial sweeteners induce glucose intolerance by altering the gut microbiota. *Nature*, v. 514, pp. 181–186.

Velasquez-Manoff, M. 2012. *An Epidemic of Absence: A New Way of Understanding Allergies and Autoimmune Diseases*. New York: Scribner, 385 pp.

West, C. E.; Jenmalm, M. C.; and Prescott, S. L. 2015. Microbiota and its role in the development of allergic disease: a wider perspective. *Clinical & Experimental Allergy*, v. 45, pp. 43–53.

Xuan, C., et al. 2014. Microbial dysbiosis is associated with human breast cancer. *PLOS ONE*, v. 9, p. e83744, doi:10.1371/journal.pone.0083744.

Acknowledgments

More than once, we asked ourselves what we were thinking after we decided to write a book together in neither of our fields. After we set out on a path with our main character—soil—we took an interesting turn and found ourselves deep in the microbial world. The unfolding story of humanity's intimate connections to the smallest known life forms thoroughly captivated us.

We would have been sunk trying to tell this story had it not been for the help of many people and we feel quite fortunate to have found an incredible team. In starting down this road, we hashed through a number of book proposals with our agent, Elizabeth Wales, who, at each junction, guided us closer to what we wanted to say and helped us find the story. She played many roles—sounding board, advocate, and muse. Her encouragement and faith in us, and the big idea underlying this book, kept us afloat at critical stages of the writing process.

Maria Guarnaschelli, our editor and champion at W. W. Norton, helped us craft our thoughts into a story with insightful questions that greatly improved the book. Maria's assistant, Sophie Duvernoy, was a joy to work with and masterful in her ability to keep us moving toward deadlines met and missed. We thank Fred Wiemer for his superlative copyediting.

Ingrid Emerick of Girl Friday Productions was instrumental in helping us wrestle unwieldy thoughts into coherent threads that preserved our individual voices while unifying our "we" voice. As every writer knows, voice is critical to the reader's experience and Ingrid coached us well. A picture truly can tell a thousand words, and Kate Sweeney converted our half-baked ideas and mock-ups into handsome illustrations.

Though this book is quite different than the one we originally envisioned writing, David Miller (now at Island Press) worked with us to gather our initial thoughts about how to use our garden to tell a story. His insights and advice were valuable and helped us test out ways of writing together and practice weaving history, sciences, and memoir together.

The number and diversity of fields that bear on microbiome research still astonish us. We are indebted to various subject-matter experts who read either several chapters, key passages, or short excerpts. Their astute corrections, comments, and clarifications sharpened our thinking and tidied up our writing.

Specifically, we thank Moselio Schaechter for generously sharing his sage perspective on and knowledge about all things microbial. A cadre of immunologists—Kristin Anderson, William Clark, Elizabeth Gray, and Amy Stone—was critical in helping us decipher and understand the fascinating and complex world of the human immune system. No less can be said for Lisa Hannon, who broadened our knowledge about essential aspects of the incredible world of plant and rhizosphere biology. Doug Fowler helped us navigate a few critical genomic details. Roger Buick enlightened us about a few important details in the early history of life. We thank Howie Frumkin for introducing us to Wes Van Voorhis, who clarified several key points about infectious diseases for us. Naturally, we alone are responsible for any errors that crept through to the final pages.

The road to understanding who we are and where we came from relies on discovery and communication. Though far too numerous to name here, this book would not have been possible without the many scientists, researchers, and authors, both today and through history, whose work we drew upon. We are grateful to have stood on such broad shoulders.

A number of friends read and commented on excerpts, draft chapters, and whole-book drafts. As any writer knows, it helps immensely to have opportunities to talk about your subject, characters, and story arc. In particular, Anne's writing-group buddies, Elizabeth Fowler and Jack Hilovsky, read drafts of material (sometimes again and again!). Their thoughts, comments, edits, and encouragement helped us identify where the story floundered or soared. Longtime friend and cross-disciplinary author and thinker extraordinaire Ann Thorpe watched the evolution of the book. Along the way, she repeatedly offered valuable suggestions for how to knit complex ideas together on the page. Polly Freeman, Sarah Ogier, and Katy Vanderpool critiqued and discussed ideas for titles and covers, as only true friends can—with honesty and insight.

Between us we have some great reading pals and they too were part of this book. Anne thanks her Food book club for their interest in and opinions on all things edible, especially June Jo Lee and Carla Levesque, who read draft material about the food choices available in America today. Anne's Baker Street book club, always a fountain of friendship and opinions, discussed many title ideas, chimed in on the cover, and accommodated her unexcused absences while we finished writing the book. And our Green book club took on the task of reading a draft manuscript. Big thinkers and doers all, their comments and discussion enriched the book and continue to enrich us. Fellow writer Gail Boyer Hayes, in particular, provided very helpful detailed comments.

We thank Dave's colleagues and students in the Department of Earth and Space Sciences at the University of Washington for their support and interest in the book. They also graciously allowed Anne to claim the table in the elevator foyer on the third floor of Johnson Hall and pile it with papers, Post-its, books, and tea mugs during several intense stints of writing and research. Brian Collins in particular was more helpful than he realized. Without reading a word, his daily encouragement of "It'll be awesome!" during the last weeks of manuscript preparation lifted the spirits of a pair of bedraggled authors. Will Brinton of Woods End Laboratories in Mt. Vernon, Maine, graciously allowed us to use a modified version of his figure of tomato roots grown with different types of fertilizer.

As this book is intended for a general audience, we have skipped an academic citation style and extensive footnotes, instead compiling the list of sources we drew upon for each chapter at the end of the book. The subjects explored in this book are far richer than can be captured in a single work, and we encourage the curious reader to explore further. But beware—like us you may quickly find yourself sucked into the perplexing questions of who you really are, how your body works, and what truly underlies fertile soil.

In the end, we remain profoundly impressed with the changes in our garden and in our own health that arose when we started to notice, pay attention to, and work with nature's hidden half. And so, finally, we want to thank our trillions of secret silent partners who keep things running behind the scenes, inside of us, and beneath our feet.

Index

Page numbers in *italics* refer to illustrations.

ABOUT THE AUTHORS

David R. Montgomery is a MacArthur Fellow and professor of geomorphology at the University of Washington. He is an internationally recognized geologist who studies landscape evolution and the effects of geological processes on ecological systems and human societies. Author of three award-winning popular-science books, he has been featured in documentary films, network and cable news, and on a wide variety of TV and radio programs, including *NOVA*, *PBS NewsHour*, *Fox & Friends*, and *All Things Considered*. When not writing or doing geology, he plays guitar in the band Big Dirt.

Anne Biklé is a biologist whose wide-ranging interests have led her into watershed restoration, environmental planning, and public health. An engaging speaker on public health and the built and natural environments, she has also worked extensively with community groups and nonprofit organizations on environmental stewardship and urban livability projects. This is her first book. She spends her free time out in the garden with her hands on plants and dirt.

Married, they live in Seattle, with their black Lab guide-dog dropout Loki.

ABOUT THE AUTHORS

David R. Montgomery is a MacArthur Fellow and professor of geomorphology at the University of Washington. He is an internationally recognized geologist who studies landscape evolution and the effects of geological processes on ecological systems and human societies. Author of three award-winning popular-science books, he has been featured in documentary films, network and cable news, and on a wide variety of TV and radio programs, including *NOVA*, *PBS NewsHour*, *Fox & Friends*, and *All Things Considered*. When not writing or doing geology, he plays guitar in the band Big Dirt.

Anne Biklé is a biologist whose wide-ranging interests have led her into watershed restoration, environmental planning, and public health. An engaging speaker on public health and the built and natural environments, she has also worked extensively with community groups and nonprofit organizations on environmental stewardship and urban livability projects. This is her first book. She spends her free time out in the garden with her hands on plants and dirt.

Married, they live in Seattle, with their black Lab guide-dog dropout Loki.